CALCULATING CURVES

The Mathematics, History, and Aesthetic Appeal of T. H. Gronwall's Nomographic Work

Ron Doerfler, Commentary
Alan Gluchoff, History
Scott Guthery, Bibliography
Paul Hamburg, Translation

DOCENT PRESS
Boston, Massachusetts, USA
www.docentpress.com

Docent Press publishes monographs and translations in the history of mathematics and computing for thoughtful reading by professionals, amateurs and the general public.

Cover design by Brenda Riddell, Graphic Details, Portsmouth, New Hampshire.

© Docent Press 2012

All rights reserved. No part of this book may be reproduced or utilized in any form or by any means, electronic or mechanical, including photocopying and recording, or by any information storage and retrieval system, without permission in writing from the author.

Contents

List of Figures   vii

Foreword   ix

Chapter 1.   T. H. Gronwall and the Spread of Nomography in America, 1900–1925   1
   1.1.   Introduction   1
   1.2.   The Spread of Nomography in America, 1900–1912   4
   1.3.   Thomas Hakon Gronwall in Europe and America, 1900–1912   12
   1.4.   The Spread of Nomography in America, 1913 – the War   16
   1.5.   T. H. Gronwall: From Chicago to the War   22
   1.6.   Nomography from the War to 1925   23
   1.7.   T. H. Gronwall at Aberdeen Proving Grounds & Washington   28
   1.8.   Conclusion   31

Chapter 2.   On Equations of Three Variables Representable by Nomograms with Collinear Points   35

Chapter 3.   Commentary   75
   3.1.   The Design of Nomograms   76
   3.2.   The Goals and Structure of Gronwall's Paper   85
   3.3.   Tests for Rectilinear Scales   88
   3.4.   The Case of One Rectilinear and Two Curved Scales   93
   3.5.   The Case of $\frac{\partial^2 \log M}{\partial x \partial y} = 0$ 104
   3.5.1.   Case I   105
   3.5.2.   Case II   111
   3.5.3.   Case III   113
   3.5.4.   Case IV   113
   3.6.   The Case of One Curved and Two Rectilinear Scales   125
   3.7.   Clark's Nomograms   135
   3.8.   Gronwall's Contribution to Nomography   146

Bibliography – Thomas Hakon Gronwall   149

References   157

Index   161

List of Figures

1.1 A simple Pythagorean nomogram. Reproduced from Hezlet, *Nomography or the Graphic Representation of Formula*, 1913, p. 25.   2
1.2 Nomogram for fire control. Reproduced from d'Ocagne, *Traité de Nomographie*, 1921, p. 360.   24
1.3 Gronwall's nomogram for computing range and deflection.   29
3.1 Nomogram for $z = x + y$.   80
3.2 Nomogram for $z = 5x + 10y + 2xy$.   81
3.3 Nomogram with rectangular outline for $z = 5x + 10y + 2xy$.   83
3.4 Circular nomogram for $z = 5x + 10y + 2xy$.   84
3.5 Circular nomogram with optimized ranges for $z = 5x + 10y + 2xy$.   84
3.6 Nomogram for $\phi(x) + \psi(y) + \chi(z) = 0$.   90
3.7 Nomogram for $z = 2x + 4y + xy + 5$.   92
3.8 Nomogram with two curved scales and one rectilinear scale.   93
3.9 Nomogram for $z = \frac{xy}{x+y}$.   101
3.10 Nomogram for $z = \frac{xy}{x+y}$.   103
3.11 Case Iα1 for $x + y + z = 0$.   107
3.12 Case Iα2 for $x + y + z = 0$.   108
3.13 Case Iβ for $x + y + z = 0$.   110
3.14 Case Iγ1 for $x + y + z = 0$.   111
3.15 Case Iγ2 for $x + y + z = 0$.   112
3.16 Weierstrass elliptic curve nomogram (rotated) for $x + y + z = 0$.   115
3.17 Zoomorphic Nomogram (rotated).   117
3.18 Weierstrass's elliptic curve with one real root.   118
3.19 Weierstrass's elliptic curve with three real roots.   119
3.20 Weierstrass's elliptic curve with a triple real root.   119
3.21 Single curve nomogram for a Weierstrass triple real root.   121
3.22 Weierstrass's elliptic curve with a double real root.   121

- 3.23 Single curve nomogram for a Weierstrass double real root. 122
- 3.24 Single curve nomogram for a Weierstrass triple root at zero. 123
- 3.25 Nomogram of Figure 3.24 after simple projection. 124
- 3.26 Nomogram with two rectilinear scales and one curved scale. 125
- 3.27 Nomogram for the equation $x - zy + z^2 = 0$. 131
- 3.28 Circular nomogram for the equation $x - zy + z^2 = 0$. 134
- 3.29 Conical nomogram for $\tan(a+b) = \frac{\tan a + \tan b}{1 - \tan a \tan b}$. 138
- 3.30 Circular nomogram for $\tan(a+b) = \frac{\tan a + \tan b}{1 - \tan a \tan b}$, $\alpha = 45°$. 139
- 3.31 Circular nomogram for $\tan(a+b) = \frac{\tan a + \tan b}{1 - \tan a \tan b}$, $\alpha = 89.99°$. 140
- 3.32 Cuspidal nomogram for the harmonic relation. 142
- 3.33 Acnodal nomogram (rotated) for $\tan(a+b) = \frac{\tan a + \tan b}{1 - \tan a \tan b}$. 144
- 3.34 Crunodal nomogram for $xyz = 1$. 145
- 3.35 Cubic nomogram for $x + y + z = 0$. 146
- 3.36 Cubic nomogram (sheared) for $x + y + z = 0$. 147

Foreword

This book is about a paper by Thomas Hakon Gronwall, "Sur les équations entre trois variables représentables par des nomogrammes à points alignés," that appeared in Liouville's journal in 1912. The paper was at the time a landmark in the mathematical literature on nomography. While a largely ignored part of Gronwall's *œuvre*, it has recently been rediscovered by mathematicians working in linearizability.

The book begins with an essay which places the paper in the historical context of Gronwall's life and the spread of nomography in America in the first quarter of the twentieth century. There follows a new translation of Gronwall's paper from the original French with editorial corrections. Next is a chapter devoted to analyzing the mathematics, visualizing the nomograms, and providing examples of the nomographic categories delineated by the paper. The last section provides the most comprehensive bibliography of Gronwall's overall work to appear, including unpublished manuscripts not previously noted.

CHAPTER 1

T. H. Gronwall and the Spread of Nomography in America, 1900–1925

ALAN GLUCHOFF, VILLANOVA UNIVERSITY[1]

1.1. Introduction

A volume devoted exclusively to the presentation and exposition of a single mathematical paper raises some obvious questions. What is the relevance of this paper, its overall importance? How did it come about? What were its effects on the development of the topic with which it deals? What attracted its author to the material in the paper?

To answer these questions we must first be clear on the meaning of nomography, the topic with which the paper deals. Nomography, given its name by Maurice d'Ocagne in 1891, is the study and construction of graphical representations – known as nomograms – of mathematical relations, primarily for use in quick and repeated calculation. The constructions involve the study and application of various geometric ideas. The study has a history extending back to the early part of the nineteenth century, when the need for such devices became apparent in engineering projects such as the construction of the French railroads in the 1840s. For a detailed account of the subject the recently issued history by H. A. Evesham [**268**] is an excellent introduction.

Over time two basic types of nomograms arose. The first was the intersection nomogram: if we restrict ourselves to relations involving three variables, say $f(x, y, z) = 0$, then in such a graph the x and y values appear on the usual vertical and horizontal axes. If specific values are chosen for them, and the co-ordinates of the point (x, y) found, the level curve of f passing through this point is labeled with the value of z which satisfies the original relation with the given x and y. In 1884 d'Ocagne, a French engineer, introduced the alignment

---

[1]The author would like to thank Richard Levitan for his helpful comments on reading this essay, and Tom Bartlow and David Zitarelli for their multiple invitations to present this material at meetings of the Philadelphia Area Society for the History of Mathematics.

1

nomogram. This device assigns to each variable x, y, and z a scale, perhaps curved, in the Cartesian plane, constructed so that if the values of any two variables satisfying the relation $f(x, y, z) = 0$ are given and located on their scales, the third may be found by placing a straightedge on the positions of the first two; the intersection of the straightedge with the remaining scale occurs at the location of the third variable satisfying the relation. See Figure 1.1 for a simple example, where we may take $x = a$, $y = b$, and $z = c$. D'Ocagne was responsible for organizing the subject of nomogram construction into a systematic body of knowledge which was presented in several volumes, the first in 1891. His later treatises appeared in 1899 and 1921 ([**178**] and [**179**]).

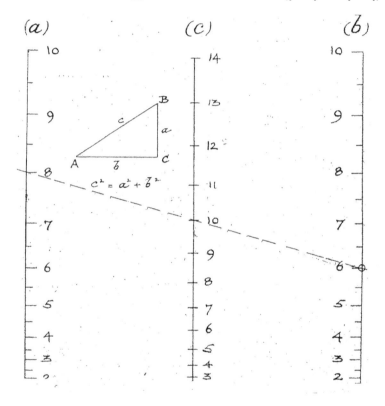

FIGURE 1.1. A simple Pythagorean nomogram. Reproduced from Hezlet, *Nomography or the Graphic Representation of Formula*, 1913, p. 25.

The alignment nomogram, in addition to its practical use, has an aesthetic appeal which is immediate on browsing the contents of any book devoted to them, especially the classics [**178**], [**179**], and [**233**]. Recently there has been a renewed interest in these beautiful diagrams; one can visit [**274**] for a particularly nice collection. There is a contrast between their mundane use and striking visual aspect; it is the hope of the contributors to this volume that it will contribute to a growing awareness of this computational device.

The paper to which this book is devoted answers, among several others, a mathematical question central to the alignment nomogram: given a relation $f(x, y, z) = 0$, what are the necessary and sufficient conditions under which it may be represented by an alignment nomogram? D'Ocagne raised this question himself in 1891, and partial answers were given by mathematicians and engineers in the following years. Much of Evesham's book is devoted to accounts of these contributions. This line of inquiry, though theoretical in nature, has the possibility of application if the answers include algorithms leading to the construction of a nomogram. Thus the problem can be considered as one useful for producing the graphs or as an interesting problem in pure mathematics. Some of the more fanciful diagrams in [**274**] have their origin in ideas which can be found in this paper.

The author of the paper, Thomas Hakon Gronwall, was a Swedish mathematician and American immigrant who published over 80 papers, but he was also a practicing civil engineer in Europe and the United States in the early part of his career as well as a professor at Princeton for a short time.[2] He was thus in an ideal position to consider such a question. His was the first work to produce a necessary and sufficient condition for the representability problem for alignment nomograms, part of the material treated in the paper to which this volume is dedicated. This was not the only way in which the problem could be answered: Evesham details four overall approaches to the problem, including Gronwall's [**268**, pp. 171–205]. But his paper demonstrates an extensive familiarity with the mathematical literature on the subject, and he relates his results to nomograms of certain kinds, such as those having straight line scales.

One can begin to see that there are several ways of looking at the paper as a product of its time, and thus begin to get some answers to the questions raised above. We can view it as one of a series of pure mathematical papers by Gronwall produced at the beginning of his mathematical career in the United States in 1912. It can be viewed as a culmination of partial contributions to the solution of a basic theoretical problem in nomography related to the alignment nomogram. In the present writer's paper [**264**] it is treated as part of Gronwall's interest in computational methods, one of his applied mathematical publications which reflect his continuing involvement in engineering and scientific problems existing in the United States at this time. Also, given that computational methods have associated theoretical questions, one may seek analogies in similar problems of the era whose theoretical underpinnings were becoming of interest. In what follows we draw the analogy to the numerical

---

[2]Gronwall's life and work in America will be discussed below; for a more complete treatment see [**264**].

integration of differential equations and the associated question of the convergence of these methods.

Finally, we may see the publication as one of several points of contact of Gronwall's career with the development of nomography in the United States during roughly the first quarter of the twentieth century. This is the view taken in the present paper. To do so one must admit to a certain bafflement: it will become clear that this development did not materially contribute to the existence of the paper, nor did the paper have any discernible effect on that development. But it is part of the quarter-century story, and since, despite a renewed interest in nomography and its history, that story is not well-known, this essay will serve a dual purpose in highlighting both the paper itself and the larger set of events of which it forms a part. One appeal of this approach is that one may get a feeling for the variety of people involved in the reception and use of nomography. This variety is a striking feature: mathematicians, engineers, scientists, machine shop workers, economists, high school students, and high school teachers were all among those who came in contact with nomograms. This account attempts to do justice to the breadth of this audience. In particular, readers familiar with the history of the American mathematical community will see certain well-known figures play a role in the story and be introduced to lesser-known but interesting ones. Readers will get a feeling for the process of the adoption of a topic in applied mathematics in the country and how it reflects the status of applied mathematics in general at that time. Section 7, which covers Gronwall's War work, contains new material on his activities there, and will serve as an example of a use of applied mathematics, including nomography, during this period.

In what follows we alternate sections devoted to the historical development of nomography in America with sections relating the career of T. H. Gronwall.

1.2. The Spread of Nomography in America, 1900–1912

Nomography was of French origin, and it quickly spread to other European nations. There were central figures in its genesis and development: d'Ocagne has already been mentioned, and we can add Rodolphe Soreau, another French engineer whose publications of a theoretical and practical nature did much to further the subject. He published a two-volume treatise, comparable to d'Ocagne's, in 1921 [**233**]. There was also a "German school" of nomography. Articles on nomography appeared frequently in such journals as the *Comptes Rendues*. America lacked such founding and guiding lights, and it may be well, before going into details, to anticipate somewhat the various viewpoints on this subject in a country which did adopt it, but without a d'Ocagne or Soreau to advocate it.

1.2. THE SPREAD OF NOMOGRAPHY IN AMERICA, 1900–1912

Above all nomography found its place first as a tool in the growth of certain heavy industries in the United States at the turn of the twentieth century, especially those involving steel structures, internal combustion engines, liquid flow in pipes, and machines of all sorts. Several engineers were responsible for writing introductory articles explaining the basic principles of the subject and showing its relevance to computations common to these settings. It was also seen as one of several "graphical methods" of computation for solving general scientific and engineering problems, methods which were becoming more popular as the American engineering community began to incorporate more European methods into their toolkits. As such it had to take its place beside methods which could involve calculating machines, tables, slide rules, or simple pencil and paper calculation; its advantages and disadvantages were discussed.

Its theoretical side gave rise to a small number of publications of a strictly mathematical character within the newly formed American mathematical community; its use in finding roots of polynomials was stressed at the college level.[3] Several prominent pure mathematicians recommended it as a subject for secondary education; teachers were exhorted to use it as a topic in the classroom, to show the "usefulness" of mathematics and interconnections among its branches. Towards the end of the twenty-five year period we are considering the nomogram became one of many methods of graphical representation of data (often called "statistics") available to workers in business, government, and social work; in these venues all graphical representations were referred to as "charts". We shall see that there was much cross-talk among these categories of use: d'Ocagne's name might be mentioned in the setting of business charts, and an engineering talk about graphical methods could take place at a mathematical congress. Nomograms as a computational tool were advocated for advancement in the machine shop setting. In contrast to Europe, the propagation of nomography in America occurred at the hands of an independent group of mathematicians, engineers, and scientists. Gronwall's career places him squarely in the events of this story.

A preamble to the story begins in Chicago, a city which recurs in later developments due both to its central location in American industry and the presence of a center of mathematical research and education located there, the University of Chicago. In 1893 the University mathematician Eliakim Hastings Moore and his colleagues organized The Chicago Mathematics Congress, an adjunct to the World's Columbian Exposition of the same year.[4] At the congress a paper on nomography by d'Ocagne was presented; in this way nomography made its American debut and was projected into the mainstream of American

---

[3]D'Ocagne's introduction of the alignment nomogram in 1884 featured an application to finding the roots of a cubic polynomial as a contrast to an earlier intersection graph for the same problem.

[4]For an account of this event see [**262**], chapter 7.

pure mathematics. As a complementary effort we may note the paper *Modern Graphical Developments* [**181**] in which the author, a local civil engineer, submitted his topic of graphical solutions to engineering problems for consideration of the mathematical audience, "in order to commend this branch of mathematics to your favorable attention – a branch which had possibly been viewed by you with somewhat less interest and attention than some more ancient and commonly cultivated branches" [**181**, p. 58]. One might easily mistake his description of the importance of graphics ("its convenience as a means of calculation in various parts of civil, mechanical, and electrical engineering and architecture" [**181**, p. 59]) for nomography. Thus we have two European topics in engineering mathematics presented at a Congress whose main purpose was as a showcase of midwestern American mathematics. This attempt to bridge the American gap between high-level mathematics and engineering was a continuing feature in the country for the first third of the twentieth century [**264**, pp. 314–317]. Although it is not known whether this was Moore's first exposure to nomography, he became one of its advocates, at least for certain purposes.

In 1900 d'Ocagne's *Traité* of 1899 was reviewed [**182**] by the geometer Frank Morley of Johns Hopkins University. While respectful of the book, Morley praises the applicability of the new discipline, stating "Practically a new subject is sprung upon us, claiming to be useful in so many directions that it would strain the faculty of an institute of technology to review the book in full detail" [**182**, p. 398]. His somewhat amusing conclusion: "The questions raised in the book should appeal not only to the technical man, but also to the teacher of elementary analytic geometry, at least to those teachers who care to heed that class of students whose cry is 'of what use is this?'" [**182**, p. 400]. The theoretical aspects of nomography are not mentioned, let alone considered of interest. His observation of the appeal of nomography to teachers of analytic geometry was elaborated upon by E. H. Moore in later years; these two research-level mathematicians were the most visible advocates, during the period considered in this section, of nomography as a secondary school subject, though neither appears to have been involved in nomography at any higher level.

At nearly the same time appeared the first exposition of the ideas of nomography to be found in an engineering journal in the United States. In 1901 an article [**183**] on the subject by Melker Johann Eichhorn was published in *Western Electrician*, the journal of the Western Electric Company.[5] The piece credits d'Ocagne's alignment nomograms with providing an efficient method for calculating mathematical relations, gives a basic explanation of the principles of the nomogram, and features charts for electrical computations and the

---

[5]The Western Electric Company was an electrical manufacturing and developing concern which grew out of America's telegraphic enterprises. The company had an office in Chicago.

determination of water condensation in steam engines. Advantages for nomogram use over table look-up are stressed; in particular it is pointed out that for mathematical relations involving four or more variables the latter is almost impossibly difficult.

Eichhorn was the first of what may be called industrial advocates of nomography, people either in industry or the academic world who worked to promote the use of nomograms in industry. Eichhorn had degrees in both mechanical and electrical engineering from Sweden and, in addition to his work as a practicing engineer, he was a charter member of the Swedish Engineers' Society of Chicago. This group, founded in 1908, was devoted to assisting immigrant Swedish technical workers in obtaining work in America. In the minutes of a meeting of the Society in 1911 [**200**] one finds a nomogram for the properties of ammonia and steam presented by Eichhorn, and the technical literature from 1911 through 1918 is peppered with charts of his construction. He also marketed a trigonometric slide rule in 1908, and designed an instrument for astronomical purposes featured in an article by an associate of the University of Chicago [**204**]. His contributions were the only American efforts cited by d'Ocagne himself [**179**, pp. xii–xiii] in the second edition of his *Traité* as having spread nomography to many disciplines.

In an echo of the 1893 debut of nomography, the 1904 St. Louis World's Fair provided another opportunity for its presentation to the mathematical community.[6] Edward Kasner, in the early stages of his strong research career, presented to the Section of Geometry of the International Congress of Arts and Sciences a paper on geometry [**185**] in which he discussed the theory of point transformations in general, summarizing the existing literature. The paper concluded with four "fields of application": cartography, the theory of elasticity, vector fields in electricity and magnetism, and nomography. Granting equal interest to both the theoretical and practical problems of nomography, he made reference to the literature "scattered through the French, Italian, and German technological journals...", and cited d'Ocagne's *Traité* of 1899.[7] With a total of two paragraphs devoted to it in roughly thirty pages of text, nomography could scarcely have been made more visible by this citation, but it indicates the beginning of Kasner's involvement with the subject, which found expression a decade or so later in graduate courses at Columbia University.

Elsewhere, E. H. Moore embraced nomography in a 1906 paper addressed to teachers of mathematics [**186**] which attempted to speak to the "fundamental problem of closer correlation of arithmetic, algebra, and geometry with one

---

[6]For the significance of the Fair and attendant mathematical activities on the growth of the American mathematical community, see [**272**].

[7]This paper was reprinted in the *Bulletin of the American Mathematical* Society in 1905.

another and with the various domains of application, or the problem of unification of pure and applied mathematics" [**186**, p. 317]. The charts were seen as a method of achieving this unification.[8] Moore also taught a course in these "graphical methods" for teachers in the summers of 1907 and 1910, backing up his enthusiasm with action.[9] The lack of involvement in the higher level theoretical problems of nomography by Moore and Kasner, however, has already been noted.

One way of documenting the growth of nomography in America is by considering the first appearance of textbooks on the subject and other evidence of courses of instruction. The first decade of the twentieth century marks the beginnings of these features. The first textbook in English which mentions nomography appears to be *Graphs and Abacuses* [**189**], published in 1907.[10] Much more visible were the efforts of Lawrence I. Hewes, a pure mathematician who received his Ph.D from Yale in 1901 and published regularly during this decade on several mathematical topics. He may be described as the first American research mathematician who devoted nontrivial attention to nomography. Beginning in 1904 he instructed a course in this subject at the Sheffield Scientific School at Yale, the notes for which eventually appeared in 1923 as [**237**]; this in addition to general courses in differential equations and planar transformations.

His enthusiasm for the subject is evident in a review of two short works by d'Ocagne [**194**]. In contrast to Morley's reserve, Hewes shows a familiarity with the earlier literature and devotes five pages to a description of the contents of the books. He is familiar with the representation problem: "It may be well to state here that the general problem, when and only when any given equation is representable by any given type or group of nomograms, is still in the main unsolved" [**194**, p. 130], but he notes partial results obtained by French mathematicians. His familiarity with projective geometry allows him to recognize that the principle of duality, which can be invoked in discussions of nomography, is a potential stumbling block for his audience: "For American engineers the idea of duality and use of parallel co-ordinates will be unattractive, but the results may be obtained entirely without use of either..." [**194**, p. 131].

---

[8]An interesting contemporary parallel is the similar work of Felix Klein, who in [**265**] looked at school mathematics from an advanced point of view, and also introduced nomograms as an application.

[9]Further information concerning Moore and his plans for reforming high school mathematics can be found in [**266**], pp. 206 ff.

[10]There is a small mystery associated with this volume: while cited in several American engineering journals and by d'Ocagne himself [**179**, p. xv], the present writer has not been able to obtain a copy of it after a thorough search of appropriate resources. It is likely that it was not widely used in America; it was published in India for use in irrigation projects there.

1.2. THE SPREAD OF NOMOGRAPHY IN AMERICA, 1900–1912

Hewes also published at least one nomogram, for use in computation of the flow of water in canals [**196**].[11]

The first surviving English-language textbook on nomography was produced in the United States: *The Construction of Graphical Charts* [**197**] by John B. Peddle of the Rose Polytechnic Institute, an engineering school located in Terre Haute, Indiana, in the industrial Midwest.[12] The book appeared in 1910, with a second edition in 1919. Peddle was a Professor of Machine Design at the Institute, and can be considered another industrial advocate for nomography. As did Eichhorn, he published a series of articles [**195**] (Peddle's appeared in the journal *American Machinist*) on the basic principles of "graphical charts" in 1908, giving many examples of their use in stress analysis of mechanical devices; he also introduced individual charts in engineering publications. His reason for undertaking this work: "...it is highly desirable that the man engaged in work requiring their use should possess some knowledge of their underlying principles in order to construct charts suited to his individual needs" [**195**, p. 753].[13] Peddle's reason for providing such a book: "Although books on nomography have been published in many foreign languages, there does not appear to have been anything written on the subject in English outside of a few scattered magazine articles... It was with the idea of supplying an elementary English text in this neglected field that the following chapters... were written" [**197**, p. v].

The level of mathematics in the book is not high, and we have an interesting contrast to Hewes' volume which brings into sharp relief the educational issues mentioned in the preceding paragraphs. While Hewes freely mentions anamorphosis and cites Lalanne, an early contributor to nomography in France, Peddle in his 1919 edition takes pains to explain and justify the use of determinants, which he has introduced in the second edition for the first time, in nomogram construction: "The word 'Determinants'... is the only drawback to the process, but I hope that no one will be deterred by the name from giving the method a

---

[11] In fact the vexing issue of the mathematical education of American engineering students began to be addressed around this time. In 1907 the Chicago section of the American Mathematical Society organized a symposium on the teaching of mathematics to this audience. Among the attendees of this event was Charles P. Steinmetz, the noted electrical engineer. Also in attendance was Edward V. Huntington, an Assistant Professor of Mathematics at Harvard University, another research mathematician who developed a modest interest in nomography, to be detailed later, as part of a much larger concern with education of engineering students at Harvard. He also was a member of an ongoing group devoted to educational issues, the Society for the Promotion of Engineering Education, which he joined in 1906. These groups of professors and professional engineers attempted to thrash out what material was necessary, suitable and understandable by their students at the college level.

[12] It is possible that H. T. Eddy, whose paper *Modern Graphical Developments* was presented at the 1893 Chicago Congress, had some connection to this school, as he was listed as "from Terre Haute".

[13] His book was based on the series of articles mentioned.

fair trial. I have used it for a number of years with classes in this subject, and my experience with them leads me to believe that anyone with the ordinary knowledge of mathematics which an engineer should possess can understand and apply the process with but little difficulty" [**224**, p. vii]. In the chapter on determinants he says: " ... if some of my statements seem to be too loosely drawn to suit the fastidious mathematician my excuse must be that I am making no attempt at a rigorous presentation of the subject, but am merely trying to furnish the novice with a useful instrument for a special purpose..." [**224**, p. 110].[14] Neither of Peddle's editions was reviewed in American mathematical journals, though it was mentioned as a New Publication in the *Bulletin of the American Mathematical Society*.

As we draw to the close of this section we should note that the spread of nomography to the branches of the American military during this era appears slow. Although the years during and after the World War saw a rapid expansion of the use of the graphs, the only mention which the present writer was able to secure relating nomography and the military is a 1908 publication [**193**] of the U.S. Naval Institute describing their use in a new method of navigation. In seeking to effect the method much computation was required, necessitating the extensive use of interpolating from tables, a time-consuming activity. Nomography provided a much quicker solution, and the author was thus led to "hope this article will be of interest to astronomers and navigators and also to naval tacticians and ordnance officers, and serve to stimulate them to work out many of their problems by *Nomography* [emphasis in article]." He confesses an "enthusiasm for *Nomography* and the recognition for solving easily and rapidly various problems in navigation, naval tactics, ballistics, target practice, etc." [**193**, p. 637]. The piece mentions Eichhorn's article and d'Ocagne's *Traité*.[15]

Following the War such uses for nomography proliferated. One can anticipate this in comments made in military journals and textbooks of the day relating to the increase in types and sophistication of weaponry. Thus the author of a 1907 textbook on ordnance and gunnery declared, "The material of war has undergone greater changes in the past thirty years than in the previous hundreds of years since the introduction of gunpowder. The weapons of attack and defense have become more numerous, more complicated, and vastly more efficient... The science of gunnery constantly requires of the officer greater

---

[14]Indeed, the subject of determinants continued to be an obstacle to some presentations of nomography even into the 1930s in America, where, for example, one has a 1939 pamphlet with the title *Mathematics of Alignment Chart Construction Without the Use of Determinants* [**248**].

[15]A Professor Pesci, to whom the author felt indebted for an introduction to nomography, himself wrote a small appendix to a book on ballistics which introduced nomographic solutions to ballistic calculations. This volume (in Italian) along with two others on ballistics were reviewed in the *Journal of the United States Artillery* [**184**] in 1903, but the notice likely raised little interest in the subject.

knowledge and higher attainments..." [**190**, preface]. The complexity and sheer numbers of the new weapons presage the amount of calculation needed to provide for their effective operation: range tables would need to be computed for any new guns, for example.

The final event related in this section is another injection, almost twenty years after the 1893 debut of nomography, of European applied mathematics into the American milieu, this time by the German mathematician Carl Runge. This took the form of a series of lectures given at Columbia University in the years 1909–1910, entitled *Graphical Methods* [**198**]. In the course Runge covered a wide variety of graphical methods for the solution of mathematical problems in differential and integral calculus, differential equations, graphical computation and root-finding, among other topics. The existence of this series is perhaps proof enough of the novelty of these ideas to American mathematicians. The present writer has argued [**270**, pp. 508–511] that, in a similar vein, there was a paucity of material related to numerical integration of differential equations (a topic mentioned in Runge's book) available to mathematicians at this time. There was again no-one in the American mathematical community with Runge's stature with regards to graphical and numerical methods, which was comparable to d'Ocagne's in nomography. The lectures contain sections on nomography (pp. 59–61, pp. 84 ff.), providing another opportunity to expose mathematicians to the ideas. A reprinted version was issued in 1912 [**201**].

Thus the status of nomography in America up to 1912 was perhaps what one would expect for the diffusion of a new foreign discipline. There were only two textbooks available for instruction of the subject (perhaps only one if we disregard [**189**]).[16] There were at least two industrial advocates of the subject (Peddle and Eichhorn). The military had limited interest in the subject. As for mathematicians, there were no published treatments of any theoretical aspects of the subject in mathematical journals by Americans at this time. Kasner, Moore, and Morley made note of the graphs, the latter two primarily recommending them for use by school teachers. Hewes and Kasner appear to have appreciated the subject for its theoretical possibilities, but Hewes alone demonstrated a high regard for both theoretical and practical uses.

We now turn to the early career of T. H. Gronwall in America and attempt to place his paper in the context of the events related.

---

[16]We have not attempted to find how many engineering schools taught this subject, though this is of course an important measure of the propagation of nomographic ideas.

1.3. Thomas Hakon Gronwall in Europe and America, 1900–1912

Thomas Hakon Gronwall was born in Sweden in 1877.[17] In 1894 he attended the Stockholms Högskola, a private institution for the study of the natural sciences and mathematics; the latter division was directed by Gösta Mittag-Leffler. Gronwall received a Bachelor of Arts from the University of Uppsala in 1896 and a Doctorate there in 1898. By this time he had already published ten papers in pure mathematics. Upon graduation Gronwall faced the common problem of a paucity of available professorships, consequently he entered the Royal Institute of Technology to increase his chances of employment.[18] He then transferred to the corresponding school in Germany, the Charlottenburg Technische Hochschule in Berlin, there obtaining a civil engineering degree in 1902.[19] He remained in Europe practicing this profession until immigrating to the United States in 1904.

From 1904 to 1910 Gronwall worked for railroad, bridge-building, and steel companies, migrating westward from Pittsburgh to Chicago:

Period 1904–1906: Carnegie Steel Company, Pittsburgh
Period 1906–1907: American Bridge Company
Period 1907–1909: Pennsylvania Lines West of Pittsburgh
Period 1909–1910: American Bridge Company, Chicago
Period 1910–1913: Consulting Engineer in Chicago.[20]

It is clear from this list that Gronwall had a lengthy exposure to the steel industry in America at this time. The American Bridge Company, as an example, was an innovator in the use of steel to build bridges and other structures.[21] These experiences resulted in two publications related to stresses on steel components [74], [79], which however did not appear until 1918 and 1919, respectively.

His status as of 1910 can be determined by examining his entry in the U. S. Census [199] of that year. He was a lodger at a boarding house with about twenty-five occupants run by a Swedish couple; many of his fellow lodgers were from Germany. He lists himself as 33 years of age, a civil engineer, working "on his own account" for the railroad, not out of work as of April 15, 1910,

---

[17]For a fuller treatment of some of the events related in this section, see [264] and [246]. Einar Hille, a Swedish mathematician who wrote the memoriam [246], was a friend of Gronwall's from 1921 until the latter's death in 1932.

[18]Eichhorn had graduated from this institution in 1892.

[19] Paul Koebe, whose work on univalent function theory began a line of inquiry to which Gronwall contributed in his years at Princeton, attended this institution from 1904 to 1905.

[20]This employment history is given in [208].

[21]One of the contemporary projects completed by the American Bridge Company was the Flatiron Building in New York City.

and single. An exhaustive search of archives shows that Gronwall was not a member of the Swedish Engineers' Society of Chicago. Perhaps he did not see the need to join since his career was already established, but it is also likely that his interests were turning away from engineering back to his true love, mathematics. In 1910 came the first indication of such a change: a letter to the Hungarian mathematician Leopold Féjèr, dated October 2, 1910, in which he discusses a certain trigonometric sum whose nonnegativity had been recently brought up in a publication by the latter. There followed a "volcanic eruption", to use Hille's phrase [**246**, p. 776], of papers, of which the nomography paper was one.[22]

We can at this point compare the lives of Eichhorn and Gronwall: both Swedish engineers attended the same technical school in Stockholm and immigrated to the United States, both resided and worked in the Chicago area in its steel-related industries, and both had nomography as a mathematical interest outside the mainstream of American engineers. But where Eichhorn used his position to advocate within industry for an acceptance of nomography as a computational tool, Gronwall was more interested in the theoretical aspects of the subject. And nomography was just part of a broader set of mathematical interests fueled by his reading at the University of Chicago, the site of his mathematical debut. On April 5 and 6, 1912, he addressed the American Mathematical Society for the first time there, presenting the papers [**14**] and [**22**].

Before discussing the nomographic results, it might be well to make a few remarks on the other papers which were part of the eruption, to see the mathematical context of which it was a part both in subject and style. In general the papers are carefully and leisurely written, some in what would today be considered great detail.[23] One moves from one set of equations to another under Gronwall's direction; the terseness of modern style of writing is absent. The paper [**15**] on which he consulted Féjèr has been described as a "masterpiece of clarity ... a pleasure to read... In our opinion, it is the most elegant treatment extant of the matter" [**261**, p. 131]. The word "elegant" comes up again in

---

[22]As an American immigrant relying at this time on industrial employment before engaging in a mathematical career Gronwall was not alone. Solomon Lefschetz, the algebraic topologist, was Russian-born and European educated, and had a work history very similar to Gronwall's. He emigrated in 1905 and worked for the Baldwin Locomotive Company in Philadelphia, then the Westinghouse Electric Company in Pittsburgh from 1907–1910 before obtaining a mathematics degree at Clark University and beginning an academic career at Princeton. Earlier, Heinrich Maschke, a German mathematician with a more varied early career in Europe, emigrated in 1891 and worked for a time at the Western Electrical Instrument Company in Newark, New Jersey, before being invited to join the mathematics faculty at the University of Chicago in 1892.

[23]The paper [**22**] contains a nearly half-page elementary explanation of Landau's growth notation and its consequences for growth calculations.

Hille's memoriam: in referring to [**33**] he notes "...the elegance of the proofs" [**246**, p. 777].

The papers cover a range of topics in pure mathematics, mostly analysis: Fourier series, the Gibbs' phenomenon, theory of numbers, the Riemann zeta-function, function theory, and Dirichlet series. They are written in English, French and German, and submitted to American and European journals. Some of these papers, like [**31**], have a computational style and purpose; the main result seems like the culmination of a single calculation. In this regard, a much later paper [**157**], on summability methods for Fourier series, has been described as "an extraordinary piece of computation" [**261**, p. 157]. Some of the results in these papers are still quoted today.[24] In some of these efforts he improves on existing results, showing a knowledge of his predecessors' work; in others he demonstrates known results by more elementary means. Hille states that "Gronwall was first and last an analyst, but he frequently went to other fields for questions to turn into analytical problems" [**246**, p. 780]. This is perhaps an apt description of his work in number theory, for example, though the subject had been blended with analysis long before. The earliest date on these papers is April 12, 1911; they list the author as from Chicago, Ill., U. S. A.

Gronwall's nomography paper exhibits many of the above characteristics. In the introductory section the goals of the paper are stated, and tribute is paid to d'Ocagne.[25] The main result, arrived at ten pages into the paper, is the necessary and sufficient condition for an expression of the form $f(x, y, z) = 0$ to be representable by an alignment nomogram. This condition is that there be a common solution to a pair of quite complicated partial differential equations. The argument is helped by the constant guidance of which particular equations are used to derive others. The equivalence proof takes up the entire first section. Gronwall indeed here went to another field for material to turn into an analytical problem. Though he traveled a known path to do so [**268**, pp. 18 ff., pp. 78 ff.], his mastery of analysis shows clearly in this complete solution. In the remaining sections he considers necessary and sufficient conditions for a nomogram to have one or more of its scales be a straight line. This is a practical issue; clearly a nomogram is easier to construct if a scale is such a line.[26] In later sections of the paper, dealing under certain restricted circumstances with producing the component functions which parametrize the scales, a deep knowledge of

---

[24]The nonnegativity of the trigonometric sum in [**15**] is called the Féjèr-Jackson-Gronwall Inequality, the number theory result in [**22**] is occasionally called Gronwall's Theorem, the equality in [**31**] is routinely called the Gronwall Area Theorem.

[25]The paper is written in French, possibly as a gesture towards the French origins of the subject.

[26]D'Ocagne's 1893 address dealt with some similar issues, but from a different point of view.

1.3. THOMAS HAKON GRONWALL IN EUROPE AND AMERICA, 1900–1912   15

the nomographic literature is displayed. Gronwall invokes not only d'Ocagne, Soreau, and their predecessor Massau, but shows a great knowledge of the work of J. Clark, whose relatively recent (1907–1908) revamping of nomographic theory was of great mathematical depth [**268**, pp. 122–157]. Gronwall's work can be viewed as an overall attempt to put the theory of alignment nomograms into the framework of partial differential equations, perhaps the most ambitious effort of his early papers. He promised a later paper in which the common solution to the two partial differential equations would be produced, but the paper never appeared. Had this solution been found, a closer connection to actual nomogram construction would have been possible. In this respect the paper, though highly theoretical, could have had more practical value. More details and analysis of the high points can be found in [**268**, pp. 173–178], and, of course, by reading the text of this book.

Reaction to the paper in the years following its publication was mixed, partly due to its unusual nature: a purely theoretical treatment of a subject with applied origins and mostly practical applications. As a work of pure mathematics it was reviewed by *Journal Für Mathematik* in telegraphic style; the reviewer noted with understatement that "Here the [representation] question is solved; of course the result, which one would not expect to be otherwise, is by no means simple."[27] D'Ocagne, who desired that nomography should become more than just a collection of techniques for chart construction but rather a higher-level mathematical discipline, acknowledged the work in a footnote in the 1921 edition of his *Traité*, but with qualifications: "The question, of purely theoretical interest, which consists of recognizing if an arbitrary equation $F123 = 0$ is reducible to this form [an alignment nomogram] constitutes a difficult problem in analysis resolved in a most remarkable fashion by M. Gronwall..."[28] Soreau made similar remarks in his 1921 book [**233**]; he also attempted a simplification in an appendix. Hille somewhat blandly described it as "a fundamental memoir on nomography" [**246**, p. 780], without justification or elaboration. In his carefully organized listing of Gronwall's work it appears under the category of "Miscellaneous".[29]

---

[27]"Hier ist die Frage gelöst; freilich is das Resultat, wie nicht anders zu erwarten, keineswegs einfach" [**250**].

[28]"La question, d'un intérêt purement théorique qui consiste à reconnaitre si une équation $F123 = 0$ quelconque est réductible à cette forme, constitue un difficile problèm d'Analyse, résolu de la facon la plus remarquable, en 1912, par M. Gronwall..." [**179**, p. 156]. $F123 = 0$ is d'Ocagne's notation for $f(x,y,z) = 0$.

[29]In 1926 Gilbert Ames Bliss, a Professor of Mathematics at the University of Chicago and an acquaintance of Gronwall's, asked Oswald Veblen, a Professor of Mathematics at Princeton and a friend of Gronwall's, for a list of the latter's "best [published] work". Veblen replied with a one-and-a-half page summary, in which the nomography paper is noted. The paper appears in the second of seven categories, namely "Nomography". "Applied Mathematics" is another category [**244**].

In America there was swift reaction in the form of an alternative solution given to the representation problem in 1915 by the American mathematician Oliver D. Kellogg, then a Professor of Mathematics at the University of Missouri, Columbia [**209**]. Kellogg's stated aim was to provide a simpler criteria for the representation problem. Though Gronwall's work is not mentioned, the paper can be seen as a response to the difficult analysis he provided.[30] Roughly speaking, Kellogg provided a necessary and sufficient condition which involved partial differentiation of f (where $f(x, y, z) = 0$ is the relation to be graphed) and related functions and determination of the ranks of matrices containing these partial derivatives.[31] There were no attempts at further classification of the graphs, nor were there references to earlier workers.

These two papers stood for a long time as the only complete solutions of the representation problem, but despite having been authored by Americans, they did not spur any new interest in the problem within American mathematics during the twenty-five period we are discussing.[32] No English-language textbook of our period cited either result. The only reference which the present writer has been able to find in the American mathematical or engineering literature even up to 1930 is in a Master's thesis in mechanical engineering submitted in 1928 [**219**].[33] Thus, with the exception of Kellogg's paper, it must be admitted that Gronwall's nomography paper had no influence in these domains at this time.[34]

We continue the story of nomography in America by turning to the next five-year interval of our time period.

1.4. The Spread of Nomography in America, 1913 – the War

In the next portion of the time frame under discussion nomography continued its rise in America. New textbooks appeared as more institutions began to incorporate this material in their instruction. More industrial advocates arose

---

[30]For an exposition of Kellogg's paper, see [**268**, pp. 179–183].

[31]This paper on nomography was the only one Kellogg published on the subject, and the topic seems to have held no further interest for him. His major area of research was in potential theory. Unlike Gronwall, it does not appear that he was involved in activities in which nomograms could be used for computation.

[32]It was not until 1959 that other equivalent conditions for representability were obtained, in the work of the Polish mathematician Warmus and the Russian mathematician Džems-Levi; see [**268**, pp. 184–204].

[33]Here the author reproduces Gronwall's two partial differential equations only as a display of the general complexity of the representation problem before turning to a discussion of simpler types of nomograms.

[34]Kellogg's paper was not mentioned either by d'Ocagne or Soreau.

to push the use of the graphs. More research-level mathematicians were attracted to the subject, and did their part to help popularize it. Nomography was present as a topic in the new engineering handbooks and shop manuals, and began to work its way into scientific laboratories. The variety of users broadened during this period.

The British military played a role in the next English-language textbook to arrive on the scene: in 1913 a nomography book by Captain R. K. Hezlet of the Royal Artillery at Woolwich was issued [**203**]. The author followed the treatment given by d'Ocagne in his *Traité*, and the immediate introduction of homographic transformations, anamorphosis, and parallel co-ordinate systems must have made quite a contrast to Peddle's work.[35] For our purposes an interesting feature of its reception in America is that it was the first English-language text on nomography to be reviewed by a journal whose contents routinely included mathematical book reviews. The *Journal of the United States Artillery* favorably received Hezlet's book, and recommended the use of nomography in general as a worthwhile endeavor.[36]

Nomography was of interest to the military as an engineering tool, and it is a sign of the increased use among its civilian counterparts that the subject began to show up in handbooks for engineers, which were themselves making an appearance for the first time. Its inclusion was due in part to Edward V. Huntington, a mathematician at Harvard University.[37] Huntington was typical of the type of mathematician which nomography attracted at this time in America: a capable researcher who by interest or need was drawn to the subject. In Huntington's case, the research interest was postulate theory, whose study in America was promoted by E. H. Moore. Huntington vigorously pursued this interest from 1901–1906 while teaching mathematics at the Lawrence Scientific School, Harvard's division of engineering.[38] Thus his position was analogous to Hewes' at the Sheffield School. When Lawrence was discontinued in 1906, Huntington became a member of the mathematics faculty proper, assuming the role of teaching the engineering students. He also became involved with general engineering educational issues, as noted in an earlier section. These connections presumably gave rise to a 1914 paper of which he was the co-author [**207**]; the paper used intersection nomograms to solve a problem in mining faults.

In contrast to this small collaborative effort we have Huntington's contributions to the massive collaboration which was the *Mechanical Engineers' Handbook* of 1916 [**211**]. The *Handbook* was meant to address the phenomenon

---

[35]More information on this volume may be found in [**268**, pp. 110–112].

[36]See [**270**] for the role played by this "progressive" journal, founded in 1892, in the exposition of new mathematical ideas in the military.

[37]The following information comes from [**273**].

[38]This interest continued until his retirement in the early 1940s.

of the explosion of knowledge necessary to practice in the profession: "It is no longer possible for a single individual to have so intimate an acquaintance with any major division of engineering as is necessary if critical judgment is to be exercised in the statement of current practice and the selection of engineering data. Only by the co-operation of a considerable number of specialists is it possible to obtain the desirable degree of reliability. This *Handbook* represents the work of 50 specialists" [**211**, p. ix].[39] As one of these specialists Huntington contributed roughly 70 pages of general mathematical tables and an eight page section on nomography, in which the works of d'Ocagne, Runge, and Peddle are referenced. Sections 1 and 2 of this work were reprinted separately [**215**] in 1918, under Huntington's name.[40] Nomography was consequently listed as part of the new onslaught of material but also as a means of calculating quantities which were part of these topics.

Similar collections of information relevant to machine shop employees without the benefit of higher education also were published at this time, offering nomograms to a new class of industrial workers. Thus in 1913 a handbook for machine designers and draftsmen came out [**202**] in which contributions from technical journals considered relevant to shop practice were "rescued from the oblivion of the out of print" [**202**, p. v]. Peddle supplied material on nomograms for this volume. Nomograms also appeared in a later book of this nature [**229**], where the impetus to learning the relevant mathematics was given a pointed rationale: "... sometimes in connection with special work, the machinist or toolmaker often finds it desirable to solve his own problems; and even though mathematics is not applied directly to the work of the shop, a knowledge of this subject will usually greatly assist the man who desires to advance. In fact, many excellent designers as well as foremen and superintendents are shop graduates who studied mathematics. This book, therefore, is intended not only to assist in the solution of the problems liable to arise in everyday shop practice, but to lay the foundation for a higher position in manufacturing and engineering practice" [**229**, pp. v–vi]. The book has a section on intersection nomograms, though they are not cited as such, and the relative merits of diagram and tabular forms of data are discussed.

---

[39]One may compare this assessment with the following comment from the military milieu: In an 1892 review of a West Point textbook appearing the previous year, the reviewer writes "When Benton's [the author's] ordnance and gunnery was first published, its excellence as a text-book was at once established. For many years the growth of artillery science had been relatively slow... Under these conditions it was possible for a single text-book to present both fully and accurately, that state of military science ... To-day, however, it is for obvious reasons, a task of no ordinary difficulty to write a text-book of ordnance. ... We cannot reasonably expect to-day any textbook to fill the place once occupied by Benton"[**180**].

[40]Based on his publication list, at least, Huntington evinced no interest in the theoretical questions of nomography.

1.4. THE SPREAD OF NOMOGRAPHY IN AMERICA, 1913 – THE WAR

In addition to the continuing efforts of Peddle and Eichhorn, nomograms by new industrial advocates were added to the literature. A 1915 presentation by R. C. Strachan [**210**] to the American Society of Civil Engineers brought the alignment nomogram to the attention of this group, in a form which should now be familiar: "This principle [the alignment nomogram] has not received from American engineers that degree of attention to which it is entitled by reason of its great value and wide range of applicability, and the object in presenting the paper is to call attention to this somewhat neglected labor saving device" [**210**, pp. 1359–1360]. Roughly thirty pages of material are presented, including charts for some contemporary engineering calculations. Peddle's book is mentioned, as are a number of other expository articles dating back to 1908.

But the most interesting aspect of this paper is the discussion among a half dozen men who critique the article, a debate which is almost as long as the article itself. One gets a nice cross-sectional view of opinions on and uses of the charts. Some of the comments offered vary from breaking entrenched habits of computation by slide-rule and tables to the accuracy of the diagrams, to the complexity of the relations which can be expressed nomographically, and the combination of graphs into one to save labor. Strachan's paper offered an opportunity for debate which was absent in earlier treatments. Again the pioneering work of d'Ocagne is mentioned several times.[41]

In contrast to the expanding industrial interest, there remains a solitary pure mathematical paper during this period dealing with nomography, an article by Hewes from 1917 [**212**] in the *Annals of Mathematics*, the Princeton-based journal. In it he introduces a "principle of adjustment" to expand the class of equations which are graphically representable by an alignment nomogram. This principle leads to the construction of systems of scales which are movable. Application is made to graphing an equation, considered by d'Ocagne, which is not reducible to a single standard alignment diagram. The principle is also applied to solve the general quartic, continuing the long tradition of nomographical polynomial root-finding which began in the 1840s. Taken together with his earlier presentation of a nomogram for water speed in a canal, Hewes again stands as the only pure mathematician of this era other than Gronwall to be involved in both the practice and theory of nomography, although at a level much lower than Gronwall and Kellogg.

During these years, however, we also have the expansion of use of the graphs from the domains of mathematics and engineering to science with the appearance of *A Manual of Chemical Nomography* [**214**] in 1918. The author, an associate in chemistry at the University of Illinois, designed a series of scales,

---

[41]There must have been a regular supply of such material by this time. The *Engineering Index Annual for the American Society for Mechanical Engineers* of the same year had Nomography listed as a category whose literature for the year could be summarized.

called "nomons", and wrote an accompanying manual to explain their use in routine calculations of the analytical chemist. He notes that the literature on such graphical computations is scarce, though he cites Peddle, Runge, and Strachan, and that "Chemical writers have generally appeared to be unaware of the remarkable recent progress in Nomography, and have based their solutions on the meager inspiration to be drawn from ordinary Analytical Geometry" [**214**, p. 5]. Reference is made to the uses of the nomons by the commercial chemist. A review of this book [**218**] brings out a possible reason why so long a period elapsed between the appearance of d'Ocagne's *Traité* and its use for chemical calculations in America, a reason no doubt applicable to other areas as well: "Chemists are naturally chary of using devices of which the underlying principles are not made clear to them. And although most of the nomograms which are likely to be needed in chemical work can be explained by simple geometrical reasoning, the proofs given by d'Ocagne and his followers are in general so abstruse in form as to be beyond the grasp of the average chemist." He excepts Peddle's book from this criticism.[42]

To round out this section we consider one last versatile individual, Joseph Lipka, in whom we have a figure who managed in his short life to combine many of the roles of the men of nomography whom we have already encountered. A Polish immigrant, he graduated from Columbia University in 1905 and received a Master's degree there in 1906. He arrived at the Massachusetts Institute of Technology (MIT) in 1908 as an instructor of mathematics; it was there that he spent the rest of his career. While there, he earned a Doctorate in 1912 at Columbia under Edward Kasner, whose nomographic activities have already been mentioned. During the summer of 1913 he traveled to Edinburgh where he met Edmund Whittaker, the creator of the Mathematical Laboratory, a setting in which both graphical and mechanical means of computation were explored and used. Lipka greatly admired the Laboratory, and consequently created a similar one at his home institution; the organizational meeting was held in in 1914. He instructed the associated course for many years, and in 1918 published a book [**216**] which summarized his course material. The volume was divided into two sections, of which the first was devoted exclusively to nomography. This exposition featured the alignment nomogram, introduced in the "hope that the simple mathematical treatment employed ... will serve to make the engineering profession more widely acquainted with this time and labor saving device" [**216**, p. iii].

---

[42]This scientific use of nomography must be distinguished from the use of the nomogram to graphically display complicated scientific laws or systems, without the primary goal of computation. Thus in [**263**] we have an account of the Harvard physiologist L. J. Henderson who in 1928 devised a nomogram to describe mammalian blood as a chemical system, but the graph was not exclusively to be used for computational purposes.

The volume had an excellent reception. This can be inferred from the favorable reviews it garnered; the large number of reviews itself is a first among such texts. According to his colleague C. L. E. Moore, Lipka himself considered the book "his most important contribution" [**240**], and Norbert Wiener, another colleague, offered in 1924 his opinion that it "constitutes one of the most valuable contributions [of graphical and numerical methods of computation] in the English language" [**241**, p. 63]. It was the first English-language nomography book to be reviewed by the *American Mathematical Monthly* [**220**], the mathematical journal devoted to university instructors and students. Though the review consisted only of quotes from the work, in a footnote is found a nomographic bibliography which gives the impression of making up for a lack of acknowledgment of the discipline by the journal. The *Coast Artillery Journal* (the new name for the *Journal of the United States Artillery*) also weighed in [**236**], as did at least a half-dozen technical journals, one containing a review by Strachan. This volume thus had wide influence, and Lipka's course inaugurated a tradition of nomography instruction at MIT which lasted into the 1960s.

But to this role of instructor and textbook author, parallel to that of Hewes and Peddle, we can add others. Like E. H. Moore, he took an interest in the topic as one suitable for secondary school students, as evidenced by a publication in *The Mathematics Teacher* in 1921 [**231**]. Unique among the mathematicians in this study, he was an industrial advocate as well: in the article just mentioned he notes proudly that "Today, some of our manufacturers are becoming interested in these charts, and the 'Department of Industrial Cooperation and Research' at the Massachusetts Institute of Technology, which is in close contact with over two hundred of these firms, has received many requests for alignment solutions of various simple problems which have arisen in their shop work" [**231**, p. 171]. He was, however, an advocate who worked from the academic side, not the industrial side. As did Huntington, he co-authored a mathematics engineering manual [**213**]. And he was responsible for the first collection of nomograms to appear in America; the volume was published posthumously in 1924 [**239**].[43] These activities were part of a teaching life which included advanced mathematical courses and a research program in geometry. But again, like the others, interest in the purely mathematical aspects of nomography was absent.

It remained only for the engineering consequences associated with the World War and a few other developments to help nomography reach the final audience considered in this paper.

---

[43]In this he predated Eichhorn, who released a similar collection, with a Chicago publisher, in 1927. Though listed as part of the collection of the University of Chicago, as of the present writing Eichhorn's volume is lost.

1.5. T. H. Gronwall: From Chicago to the War

After his auspicious debut at the AMS meeting in Chicago Gronwall accepted a position in the mathematics department at Princeton University, beginning in the Fall of 1913. His research was impressive, but his background in engineering was also key to his promotion in 1914.[44] His accomplishments there are related in [**264**, p. 323], as are the circumstances surrounding his dismissal at the end of Spring 1915. He then entered a period of personal freefall, which ended only in an invitation to join the War work at the Aberdeen Proving Grounds in 1918. Yet he still managed to produce publications.

But this period is also of interest because it threw him back on his engineering work, albeit in an irregular series of positions. Some of these jobs were in New York City, to which he relocated in 1916. Perhaps these contacts moved him to address the American Mathematical Society on February 26, 1916, presenting the contents of his papers on stress in cylindrical keyways and shafts, interests dating from his Midwestern engineering days. As mentioned in [**264**], these papers were noteworthy for their reduction of the solutions to a form suitable for numerical calculation, a step not always taken in similar mathematical stress analyses. He also embarked on two pure mathematical projects, each extending to his time at Aberdeen and subsequent position on the ballistics Technical Staff in Washington. These efforts may have had their origin in an attempt to improve his financial situation from book sales. The first was a translation from the Danish and an expansion of a lengthy article by J. L. W. V. Jensen on the Gamma function [**70**], followed by his own treatment of this function [**75**].[45] These works were published originally in the *Annals of Mathematics*, but were later issued as reprints for sale by that journal for 50 cents and 90 cents respectively. The other project was a planned book on univalent function theory which never came to fruition, though portions apparently were published as separate notes [**264**, pp. 329–330].[46]

---

[44]According to the Faculty Minutes from September 24, 1914, "...his work [is] to be primarily in the department of Civil Engineering [Strachan's profession]. Professor Gronwall comes to us with a record for fine scholarship both on the Continent and in our own country...[he] will be of great assistance in the problems of Mathematics with the Engineering students, a point in our instruction where we have been unfortunately lacking" [**205**, p. 29].

[45]In the latter work he states in the introduction that "While the paper contains little that is new in subject matter, a considerable number of proofs have been remodeled, or replaced by new ones." This is the only such expository work done by Gronwall. Several modern-day college libraries still report ownership of this volume.

[46]Had he been able to remain at Princeton Gronwall was planning to give a course in this material in the Fall of 1915. Some of the funding for this latter work was administered by E. H. Moore; a series of letters [**277**] between him and Gronwall describes the progress and reversals of this project.

Gronwall's activities during this period also involved reviewing mathematical literature, as one can see by the listings in this volume. Much of this reviewing was done for the *Bulletin of the American Mathematical Society*. One such review was for the 1915 German language version of the Runge book on graphical methods, which grew out of Runge's lectures at Columbia University.[47] His notice is basically a list of the contents, but two comments are interesting. He cites as part of the material in the second chapter "the calculation of $z = f(x,y)$ by contour lines of the corresponding surface, and its dual method in line coordinates, the nomography of d'Ocagne" [**61**, p. 408]. This is the mathematician's characterization of the alignment nomogram using the terse nomenclature of projective geometry.[48] The article concludes with the remark: "The book under review brings forth one sad reflection: when will our writers of calculus texts for engineering students see fit to give something really modern and practical on graphical integration and solution of differential equations?" Here Gronwall is appealing for a college-level presentation of an engineering topic in the same category as nomography, for which only Peddle's and Hezlet's books were available. His time at Princeton must have brought this deficiency to his attention. Thus we see that this period shows Gronwall's activities as researcher, instructor, and engineer, all occupations relevant to nomography.

1.6. Nomography from the War to 1925

In a 1921 publication[**231**, p. 171] Lipka claimed that "The world war brought our ordinance engineers in contact with the French engineers, and the former have learned how the latter apply the principles underlying the alignment chart to the graphical solution of some of their problems in ordnance." Details of such meetings are lacking, however; were they to become available, they would provide interesting evidence for the direct influence of the creators of nomography on American usage. There are references in the contemporary American ballistics literature linking the French with "graphical methods".[49] The 1921 edition of d'Ocagne's *Traité* and Soreau's corresponding volumes show many examples of the French use of nomograms for fire control. A nomogram for differential corrections to a trajectory, for example, can be found in the 1921 *Traité*, see Figure 1.2.[50]

---

[47]This series was mentioned in Section 2 of the present paper as a high-level presentation of graphical methods made by a European mathematician in America in the period 1900–1912.

[48]He also notes the lack of mention of any works of d'Ocagne for further study.

[49]One such reference occurs in a post-War survey of European methods of ballistic theory written by Oswald Veblen after an information-gathering Contintental trip [**225**, p. 10].

[50]This topic of differential variations is one to which Gronwall devoted much time, as will be discussed below in Section 7.

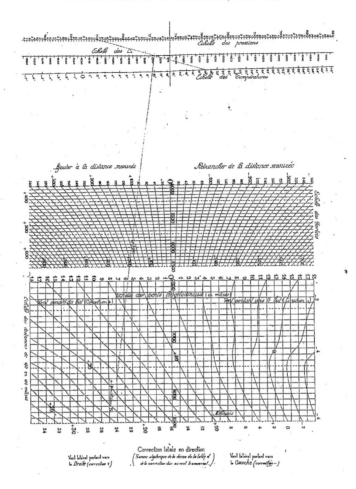

FIGURE 1.2. Nomogram for fire control. Reproduced from d'Ocagne, *Traité de Nomographie*, 1921, p. 360.

The War provided a new outlet for the growing discipline in the United States as well. One has only to look, for example, at the key summary of a year-long stretch of activity at the Ballistics Branch of the Ordnance Department in Washington, D. C. [**226**], to see its presence. Organized by Forest Ray Moulton of the University of Chicago, this was one of several groups of mathematicians and engineers working on ballistics problems at the time; the report summarizes the activities from April 6, 1918 to April 2, 1919.[51] On pages 34–35 are listed twenty-nine "tables, diagrams, and nomograms" which were used to ease the

---

[51]For further information about this group and that of Oswald Veblen at the Aberdeen Proving Grounds, see [**264**] and [**270**]. The latter contains a discussion of the pre- and post-War history of the numerical integration method of Moulton.

1.6. NOMOGRAPHY FROM THE WAR TO 1925

burden of computation for the newly instituted numerical method of trajectory computation, introduced by Moulton himself. Document 80 in this list, for example, by Moulton, details the construction of nomographic curves for the calculation of certain key functions in his method: "The purpose of this paper is to show how the use of logarithms and computing machines may be entirely avoided and the labor very much reduced by employing two families of curves, which have been constructed and which accompany this discussion" [**223**, p. 1]. Trajectory computations by Moulton's method involved great amount of calculation, and in this case the nomograms were introduced, perhaps in a novel way, to aid a numerical method for integrating differential equations.

The diagrams must have found many other such applications; support for their use can be found in a memorandum on range computing [**264**, p. 320] issued by the Aberdeen Proving Grounds in 1919. In that same year the *Journal of the United States Artillery* announced a summary article on nomography, which did not appear until 1923 [**238**]. According to the author, who also had prior experience in working with the new numerical integration methods, "Due to their comparatively recent origin and notwithstanding their present extensive use in gunnery as well as in engineering, the theory of the nomogram is not common knowledge even among engineers" [**238**, p. 157]; he added the by now common remark that engineers may be reluctant to change their familiar methods for performing calculations. The paper culminates in the construction of four charts for target practice, which were used at Fort Casey, Washington, in 1916.[52]

From the war emerged a small group of mathematicians displaying to varying degrees an active interest in mathematical methods of computation [**270**]. Among them was Lester J. Ford, whose writings on the numerical integration of differential equations were mentioned in the previous citation, [pp. 537–538]. Having received both Master's and Doctorate from Harvard in 1913 and 1917, respectively, under Maxime Bocher, he spent the years 1914–1916 in lecturing and doing research in Edinburgh.[53] From his instructorship at Harvard he was called in 1918 to Moulton's group, where he served for about four months. But Ford's most lasting contribution to the War effort was a seventy-one page booklet *Elementary Mathematics for Field Artillery* [**222**], part of an effort to raise the educational level of the instructors of military mathematics. The book starts at the beginning with "whole numbers" and fractions and works its way through the topics necessary for military use. Among these topics is a one-page section on "The Calculating Chart", which illustrates the use of an alignment

---

[52]Though we have not surveyed the literature, the Navy as well surely used nomograms. Evidence for this can be found in later publications in engineering journals by individuals associated with naval positions.

[53]It would be interesting to know if Ford, like Lipka before him, had any contact with Whittaker's Mathematical Laboratory during these years.

nomogram for addition, presumably as preparation for the use of more elaborate charts.⁵⁴ Ford maintained interest in this subject, for after his research career at the Rice Institute in Texas he relocated to the Illinois Institute of Technology, where he published in 1944 a treatment of nomography [**251**]. It shows the earmarks of a mathematician in its unapologetic use of standard mathematical terminology and avoidance of short-cuts for application.

The War effort involved huge amounts of calculation; the applications we noted describe only the ballistic sciences. This calculation in turn produced correspondingly large amounts of data as, for example, the many different types of guns and ammunition had to be accounted for in range firing tables. This data stream joined a larger flow after the War which found expression in the form of charts other than nomograms and was directed for the first time to a non-engineering audience: "In and since the War the use and development of charts has been almost phenomenal – so large in fact, that at least one able economist who is interested in such things thinks that we as a country have gone chart-mad. ...This development ...has a very solid basis in practical utility. There is little question that the chart represents a genuine saving in time and in mental effort" [**242**, p. xxxviii]. This is how the phenomenon is explained in a typical post-war volume devoted to the subject. A standard account of the evolution of this trend cites Peddle's book as the first treatment in the English language dealing with graphical representation *in general*, as opposed to just nomography. Following this path, a 1914 work *Graphic Methods for Presenting Facts* [**206**] was mentioned as the first such book requiring no mathematical background; it could be understood by "the average man of business" [**232**, p. v]. After the war "charting" gained interest as many such volumes began to appear.

The 1914 work just mentioned listed as its proposed readers a collection of non-scientific workers, people presumably addressed for the first time by the chart-makers: " ...the businessman, the social worker, and the legislator... anyone dealing with the complex facts of business or government" [**206**, p. v]. In its seventeen chapters one gets a nice cross-section of applications of graphics as the terminology of data-collection was also introduced: frequency curves, moving averages, factory production schedules, and annual reports of corporations are all mentioned as are mistakes and tips in the construction and presentation of graphs. Intersection nomograms are discussed [pp. 333–335], and alignment nomograms are favored in [**232**], in which a nod to d'Ocagne

---

⁵⁴In an assessment of the lack of military knowledge on the part of college mathematics faculty, contained in a review of this book [**221**, p. 353], Albert A. Bennett, another mathematician participating in the War effort, wrote: "Such terms as mil, grad, deflection, azimuth, dispersion, diagram, probable error, center of impact, nomogram, or alignment chart, meant little or nothing to many college instructors in mathematics..."

occurs. Although titles relating to general engineering graphic charts, including nomograms, also appeared at this time [**228**], the interest on the part of economists, managers, and bankers was stressed in the new literature. In this respect the nomograms became part of the tidal wave of information accompanying the post-war business environment where data needed to be presented with speed and factories and offices managed as efficiently as possible. The graphs were even included in some algebra courses in commercial high schools [**217**, pp. 201–208].[55]

We now turn away from these last two developments to finish up the quarter-century with some miscellaneous events. In the way of texts, Peddle's book appeared in its second edition in 1919 [**224**], Hewes' [**237**] in 1923. The latter was praised for its "sound logical mathematical background" in a review in the *Bulletin* [**245**]. In 1920 a new textbook by an English Professor of Mathematics was published [**227**]. The volume found favor in undergraduate classes and was, as Lipka's work, widely reviewed.[56] Other texts appeared, but they began to tend toward a uniformity of presentation, perhaps a sign that the subject had found its place in academia at last. As for research mathematicians, in the summer of 1919 Kasner offered a graduate course "Graphical methods including nomography and applications of the calculus" at Columbia, and in 1922 he supervised a Master's thesis devoted to nomography, the first such thesis of its kind in America [**235**]. In the introduction, where the nomogram is listed as a geometric object along with transformations in geometry effected by ruler and compass and the graph of the Cantor function, reference is made to all the users of the graphs we have already encountered, including applications in business. A six-page bibliography cuts across all disciplines, and it is evident from such a work, which has a tone of youthful enthusiasm, that nomography had established itself in its broadest way by this time. Further evidence of the presence of nomography in the mathematical undergraduate world was a steady stream of papers in the *American Mathematical Monthly* in which the graphical solutions are employed. Some of these provided additional "catching up on the subject" bibliographies, as had the earlier [**220**].

---

[55]A particularly moving photograph appears in [**242**, Fig. 238], showing a "statistical exhibit" as part of a Municipal Parade in New York City in 1913: "Many very large charts, curves and other statistical displays were mounted on [horse-drawn] wagons in such a manner that interpretation was possible from either side of the street. The Health Department, in particular, made excellent uses of graphic methods, showing in most convincing manner how the death rate is being reduced by modern methods of sanitation and nursing."

[56]The *Bulletin* and the *American Mathematical Monthly* gave summaries, though the latter again simply quoted from the contents.

1.7. T. H. Gronwall at Aberdeen Proving Grounds & Washington

Gronwall's period of wandering following his leaving Princeton came to an end when he received a call to join the War work at Aberdeen Proving Grounds in the Fall of 1918. The full range of his work there and at his later appointment in Washington has never been discussed, but it provides another opportunity to understand the nature of his mathematical work, especially with regard to computation. Indeed, his earliest paper[57] was a three-page memo describing a "small arc" method for computing trajectories, a preliminary report [**78**]. One of many such methods to follow Moulton's original approach[58], it lays out a step-by-step procedure, with advice on interval sizes and other matters. A trajectory is computed for the first several arcs. Although apparently never followed up, this paper shows Gronwall to be one of the few American mathematicians familiar with numerical integration at this time. Another report [**91**] offered an alternative treatment of the Bliss theory of differential variations, to be described below. Next completed was [**92**], an early version of [**97**], an effort which took him in a direction opposite to calculation: theoretical treatment of the properties of the ballistic trajectory. Taken together the three reports display that mix of computational knowledge and pure mathematical interest which characterized his interest in nomography: he could not only plot the trajectory and small variations from it, given the necessary data, but also deduce general mathematical consequences about it.

In fact Gronwall produced a nomogram at this time for computing range and deflection from co-ordinates [**93**]. This activity was part of the routine work at Aberdeen where test firings were repeatedly performed, making them a clear candidate for calculation by nomogram. The graph was used to calculate the deflection from the plane of fire and range of a projectile given the co-ordinates of the gun and the "splash" (the point of fall in the Chesapeake Bay). Directions for use are carefully given as are two examples; then a section is devoted to the "theory" of the chart. Last are given "practical hints" for drawing it; this involved knowledge of the drafting tools of the day and availability of celluloid slide rule scales to mark the graph's scales. The final nomogram as drawn by Gronwall is shown in Figure 1.3, complete with draftsman's lettering, signed and dated May 25, 1919.[59]

---

[57]The following papers produced by Gronwall at Aberdeen can be found in Record Group of the Department of Ordnance, Record Group 156, Entry 866, National Archives and Record Administration, College Park, Maryland.

[58]For a discussion of these see [**270**].

[59]A later note [**95**] described a modification of its use when deflection was computed from target records; this involved linear deflection rather than the angular deflection of the original chart.

1.7. T. H. GRONWALL AT ABERDEEN PROVING GROUNDS & WASHINGTON

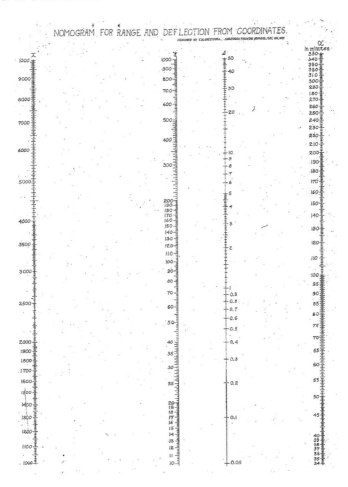

FIGURE 1.3. Gronwall's nomogram for computing range and deflection.

Gronwall's last effort at Aberdeen [94] was an exercise in cartography, and as such shares features with nomography (recall Kasner's 1904 address in St. Louis which listed cartography and nomography as applications of point transformations in geometry.) The goal can be most easily grasped by quoting the introduction: "In locating the point of fall of a projectile by means of azimuth observations from fixed points (towers), and the subsequent determination of range and deflection from these data, it is necessary to take the curvature of the earth into account, even for short ranges, as soon as the mutual distances of the towers are large. This object may be accomplished by mapping the earth spheroid on the plane in such a manner that the azimuth lines on the spheroid are transformed (with sufficient accuracy) into straight lines in the plane; the determination of the range and deflection is then achieved by the ordinary formulas of plane coordinate geometry." The report thus describes a transformation in which certain curves are changed into straight lines. This kind of change is routine in constructing nomograms. The document follows Gronwall's established style: variables are laid out, transforms are defined, accuracy is discussed, examples are computed, page after page of trigonometric formulae and series expansions follow, suggestions for computational tools are made, and finally details on constructing the map itself are given. It forms a part of the activity surrounding the computation of range and deflection for firings for which his nomogram was drawn.

In mid-July of 1919 Gronwall transferred to the Office of Ordnance in Washington, D. C., where he remained as a member of the Technical Staff until 1922. Here his work was almost exclusively devoted to understanding and computing the differential variations of a trajectory. This topic was a traditional part of ballistic theory, but a new, more mathematically advanced approach was introduced by Moulton and later developed by Gilbert Ames Bliss, a student of Moulton's at the University of Chicago also called to Aberdeen. The variations in question were those in the range of the projectile introduced by relatively minor factors: cross winds and following winds, changes in temperature and density of the atmosphere, small changes in the elevation of the gun. Bliss' treatment of these perturbations was rapidly becoming the more popular by the time Gronwall came on the scene. Gronwall almost immediately was able to introduce a mathematical simplification in the theory which both reduced the amount of labor involved in computation and allowed for theoretical understanding of the system of differential equations generating them.[60] In [**96**][61] he wrote a comprehensive account of an extension of Bliss' method which considered effects of the variations on time of flight and maximum ordinate. Short individual memoranda on the effects of changes in gravity and rotation of the earth were issued as [**103**] and [**109**]. His other works from this post can be gleaned from a 1922 list of "mathematical papers" by members of the Technical Staff which sets aside ten slots ([**110**], [**111**], [**112**], [**116**] – [**122**]) for works of Gronwall. Some of these were tables for which he did the computational work.

Although another theoretical ballistics paper [**104**] emerged from this setting, and other non-ballistics papers were published, the labor of the computational work was probably a factor in Gronwall's sudden abandonment of his position on the Technical Staff and his departure for New York City, where he remained for the last decade of his life. In this period he was based at

---

[60]See [**264**] for some details of this theory and Gronwall's contribution to it.

[61]The following papers produced by Gronwall while a member of the Technical Staff can be found in Record Group of the Technical Staff, Record Group 156, Entry 981, National Archives and Record Administration, College Park, Maryland.

Columbia University and considered many problems in physics, physical chemistry, and pure mathematics, producing in the process many tables and approximate formulae, adding to the flow of data, charts, and numbers washing over the country. He died in 1932.

1.8. Conclusion

It will be good to answer the questions raised in the Introduction before concluding this essay. Gronwall's nomography paper was the product of the effort of a pure mathematician interested in the mathematical bases of a tool with which he was familiar from his work as an engineer. He attempted to relate this tool to the larger body of computable mathematical expressions which occur both in practice and in theory. It provided a complete solution to a mathematical problem associated with this tool, the representation problem for alignment nomograms, as well as many other related results. It was not the only solution produced, as it was followed shortly by another (Kellogg's), and much later by solutions with greater degrees of applicability. The paper ambitiously attempted to classify nomograms from the point of view of partial differential equations, and thus unify the scattered partial results up to that time and give an overview of the subject. However, its effect on theoretical nomography was limited during our period to inspiring other approaches to the representation problem, and there were no attempts to build on Gronwall's work at this time. It had no effect on practical nomography. The paper stands unique as a heavy piece of classical analysis of a type similar to other works of Gronwall's mathematical debut and his later work, analysis devoted to the understanding of a deceptively simple device.

Its peculiar position has an analogy in another computational topic of the day which is worth mentioning, as it shares features with the representation problem. It is argued in [270] that the introduction of numerical methods of integration for systems of differential equations by Moulton at Aberdeen and the advocacy of the topic by certain mathematicians after the War raised the level of visibility of this technique in the mathematical and engineering communities. But Moulton was adamant that nearly anyone with a sufficient knowledge of differential equations could have effected such an introduction, for which he was receiving great praise at the time. Almost from the beginning he was equally concerned with an accompanying theoretical question: under what circumstances is such a method guaranteed to converge to the exact solution of the system? This question was not an afterthought, as he wrote a technical report containing a solution early in his War work in 1914. Different versions of this convergence proof appeared in print, none in a mathematical journal. This contribution was of little interest to the users of his method, who often found the method itself tedious. Mathematicians, with few exceptions[62], were not eager to embrace numerical integration, let alone concern themselves with convergence criteria or other associated theoretical issues. In spite of some intermediate literature [**247**], it was not until the second World War and afterwards that such matters began to be seriously examined on a large scale. Thus we see another computational subject whose theoretical aspects were not significantly developed by American mathematicians at this time.

Moulton was an astronomer familiar with modern analysis, which he called upon in his astronomical and ballistic work. Gronwall was a pure mathematician who relied on engineering work for employment. One of the tools of this occupation provided him with subject matter for the exercise in pure analysis which was his nomography paper. But his was not the common way in which American mathematicians of the day became involved with nomography, as should be evident from the sketches we have provided. Kellogg was an analyst who wrote a single paper on nomography, to which he never returned. Some, like Morley, Moore, and Kasner, were pure mathematicians whose interest was largely pedagogical. It was for them a subject suitable for introduction in high schools or graduate courses for interested students. In Moore's case, it was part of a larger ongoing secondary school mathematics reform effort. Others, like Hewes and Lipka, embraced it as an engineering topic for instruction, the first showing an interest in its deeper mathematical context, the second acting uniquely as both an industrial and educational advocate. Huntington's and Ford's careers mixed its instruction in engineering settings with their own unrelated research. The majority of these men would have been considered geometers, or at least had geometry as a field of interest. Gronwall and Kellogg were analysts.

We can round out this survey with a tenth and final portrait. Harris Mac-Neish was a high school teacher in the Chicago area who took a Master's degree at the University of Chicago in 1904 and a Doctorate there in 1909, both on topics in projective geometry under the supervision of the ubiquitous E. H. Moore. MacNeish initially took an instructor's position at Princeton but shortly thereafter accepted an appointment at Hewes' institution, the Sheffield Scientific School. He published a number of articles in the *Annals* and somehow acquired an interest in nomography. In 1925 he presented a paper at an Annual Meeting of the American Mathematical Society on *A nomogram in n-dimensional space for the solution of n linear simultaneous equations* [**243**]. The work was a generalization of a known nomogram for two equations in two variables; the generalization was achieved by a method in projective geometry presented at an earlier meeting. This 1925 work was not published. (It clearly had only limited practical use!) The combination of ample research ability with a passing

---

[62]Lester Ford was minimally interested in the convergence question; see [**270**, pp. 537–538].

interest in an application seems to characterize his effort, as well as those of several of our earlier men.

We can also reverse the perspective and ask how nomography was viewed by the American mathematical community at this time and later. For the era under consideration it is clear that it was not highly regarded, and it appears that it always remained a subject with much less than first class status, with little attention paid to it in mathematical journals over the years following the first quarter-century. It never attracted the attention it received by contemporary European workers, where figures such as d'Ocagne and Soreau could raise its visibility, let alone the intense interest of the Polish and Russian mathematicians in the 1950s and 1960s, as described in [**268**, pp. 207–244], for example. At the risk of reading too much into a remark, one gets a sense of this lower status from the introduction of a 1947 textbook on nomography written by two highly regarded instructors at MIT [**252**]: "The theory of the nomographic chart cannot be dismissed as a simple topic. Mathematicians who do so are unaware that a complete treatment of the subject draws on every aspect of analytic, descriptive, and projective geometries, the several fields of algebra, and other mathematical fields." This defense seems to be mounted against criticism of nomography as less than demanding material.[63]

All of this ignores the much larger setting of the absorption of nomograms in the cultural picture due to their labor-saving qualities, so often mentioned above. This was an age of computational activity on a mass scale, most easily exemplified by the work at Aberdeen Proving Grounds and the contents of the new engineering manuals, the latter full of numbers for use in problems of all kinds. It existed wherever stresses on beams, flow of water in a pipe, the yield of a bond, or the turning time for a piece on a lathe needed to be repeatedly calculated. It was one of many devices, including the pure mathematical work of men like Gronwall and Bliss, which were pressed into the service of reducing computation, producing data quickly in a usable form. Data processing became a matter of broader attention during and after the War, when the users of such information grew from the scientific and engineering communities to include the business, governmental, and helping profession workers. As a visual means of calculation and variable representation it was classed with pie charts, bar graphs, and frequency diagrams with which it shared this appealing advantage over tabular expressions of information.

---

[63]In this respect, it somewhat resembles the exterior ballistics of that day. Its origins are practical, some deep analysis can be brought to bear on the problems it raises, but few American mathematicians devoted more than a passing interest to it. Soreau and d'Ocagne played a role for nomography in France similar to that played by Paul Charbonnier for exterior ballistics; his volumes were considered mathematical classics on the subject. Our last mathematician, MacNeish, published an elementary book [**249**] on exterior ballistics in 1942 at his final institution, Brooklyn College.

In the mainstream of engineering and mathematics education the trends established during the first quarter-century continued. Master's theses on nomography appeared sporadically over the following decades. New texts were published.[64] As noted above, these books had for the most part a sameness of topics and presentation lacking in the pioneering efforts of our period. Courses proliferated in engineering schools as the schools themselves grew in number and size. More chart collections were published by some of these schools; charts for individual disciplines such as chemical engineering, mining, navigation, optics and construction also were printed. As new industries emerged, so did nomograms dealing with the calculations for them. Use in the military increased with greater types of weapons and new inventions such as the airplane. Their function as a unifying topic in high school mathematics lessened, though occasionally reappeared. New industrial advocates wrote about the virtues of the nomogram. One such devoted an entire bachelor's thesis [**254**] to a survey of usage of the charts in industry, complete with praises and criticisms of them by the users. In this work we find a reference to a risk inherent in any accepted method of calculation: "Throughout the rest of the paper you will see many lines in praise of the way nomography permits unskilled help to solve complicated problems. On the other hand, you will be exhorted not to use nomograms unless you have a complete understanding of the basis for the nomogram, and in effect can do the problem 'longhand' " [**254**, p. 30]. Such warnings are often given for methods used so widely and for such long periods of time that the specter of their possible adverse effects on understanding the underlying problems is raised.

But Gronwall's work lies in a different realm than all of this. In the present volume his paper is explored in detail for the first time, and examples are given which illuminate its contents. Whether regarded as a fascinating piece of classical analysis or a guide to the world of nomography and its possibilities, the contributors to this volume hope to instill an interest in this piece which it richly deserves. Additionally, the present writer, whose first knowledge of nomography came from discovering Gronwall's paper, hopes the present essay has clarified its relation to the story of nomography in the era of its introduction in America, a story from which it is separate yet paradoxically of which it also a part.

---

[64]The online OCLC database lists roughly a dozen titles on the subject from 1924 through 1940 in settings ranging from classroom use to bulletins issued at university experimental stations.

CHAPTER 2

On Equations of Three Variables Representable by Nomograms with Collinear Points

Thomas Hakon Gronwall

INTRODUCTION

Given the equation

(1) $$F(x, y, z) = 0,$$

where F is an analytic function of three variables, suppose that it is reduced, by whatever methods, to the form

(2) $$\begin{vmatrix} f_1(x) & g_1(x) & h_1(x) \\ f_2(y) & g_2(y) & h_2(y) \\ f_3(z) & g_3(z) & h_3(z) \end{vmatrix} = 0.$$

Let us designate ξ and η as the Cartesian coordinates; then (2) expresses that by taking the points corresponding to $t = x$, $t = y$ and $t = z$, respectively, on the three curves,

(3) $$\xi = \frac{f_i(t)}{h_i(t)}, \quad \eta = \frac{g_i(t)}{h_i(t)}, \quad (i = 1, 2, 3),$$

these three points will be located on a straight line. By marking the points on each of the curves (3) that correspond to the rounded values of t, we obtain three *scales* and by joining the points corresponding to the given values of x and y, for example, and reading the value of z at the place where this intersects the scale of z's, we obtain a graphical solution for equation (1).

This is the principle of M. d'Ocagne's nomograms for collinear points ; he developed a theory in this regard that is remarkable for both its analytical elegance and its practical utility, especially in the engineering profession.[1]

---

[1] M. d'Ocagne, *Traité de Nomographie*. Paris, Gauthier-Villars, 1899; *Calcul Graphique Et Nomographie*, Paris, Doin, 1908.

The fundamental problem of this theory is obviously how to recognize whether a given equation (1) can be reduced to the form (2) or not. For specific classes of equations (1), certainly including the majority of equations encountered in practice, we know how to accomplish reduction to the form (2); but for the general case, the conditions for reduction have remained unknown up to this time.[2]

In Section 1 of this study, I demonstrate that the necessary and sufficient conditions for equation (1) to be reduced to the form (2) consist of the existence of a common integral for two partial derivative equations. In Section 2, I provide additional conditions for one or several of the scales to be rectilinear, which is an important case in practice, and in Section 3, we discover a method for effectively extracting the functions f_i, g_i, h_i, from the integral in Section 1 that is presumed to be known. This method becomes illusory if two of the scales are rectilinear. In Section 4, we will first assume that all three scales are rectilinear; these cases must be dealt with separately for two reasons: on the one hand, the given equation then allows for essentially distinct nomographic representations, all of which I can determine, and on the other hand, the formulas in the following sections contain as their denominator an expression which is canceled out in this particular case. Excluding the method in Section 3, there still remains the case of two rectilinear scales and the third one being curved, which is the object of Section 5. In the sixth and final section, I specifically examine Clark's very remarkable nomograms where two of the scales are supported on the same conic, and the third scale is of whatever kind.

In a separate paper, I will explicitly construct the common integral of the partial derivative equations from Section 1, and I will show that the case of Section 4 is the only one where the given equation allows for essentially distinct nomographic representations.

1. – Necessary and Sufficient Conditions

In permutating any existing columns in (2), we can proceed in such a way that $h_1(x)$, $h_2(y)$, $h_3(z)$ are not identically canceled out; thus by dividing each line by the corresponding function h, we can assume that $h_1 = h_2 = h_3 = 1$ and (2) becomes

(4) $$\begin{vmatrix} f_1(x) & g_1(x) & 1 \\ f_2(y) & g_2(y) & 1 \\ f_3(z) & g_3(z) & 1 \end{vmatrix} = 0.$$

---

[2]Only M. Duporcq had obtained, in the form of functional equations, some sufficient conditions. (See D'OCAGNE, Traité de Nomographie, p. 427-431).

2. ON EQUATIONS OF THREE VARIABLES

Following along with d'Ocagne, let us multiply equation (4) by a determinant

$$\begin{vmatrix} a_1 & a_2 & a_3 \\ b_1 & b_2 & b_3 \\ c_1 & c_2 & c_3 \end{vmatrix} = 0;$$

after division by the elements of the third column, this becomes

(5)
$$\begin{vmatrix} \bar{f}_1(x) & \bar{g}_1(x) & 1 \\ \bar{f}_2(y) & \bar{g}_2(y) & 1 \\ \bar{f}_3(z) & \bar{g}_3(z) & 1 \end{vmatrix} = 0,$$

where

(6)
$$\begin{cases} \bar{f}_i = \dfrac{a_1 f_i + b_1 g_i + c_1}{a_3 f_i + b_3 g_i + c_3} \\ \bar{g}_i = \dfrac{a_2 f_i + b_2 g_i + c_2}{a_3 f_i + b_3 g_i + c_3} \end{cases} (i = 1, 2, 3).$$

One thus obtains an equation that is equivalent to (4) by performing the most general homographic transformation on the f_i, g_i.

After this remark, whose significance will soon become apparent, we then develop (4) according to the last line and obtain

(7) $$g_3(z) = u f_3(z) + v,$$

where

(8)
$$\begin{cases} u = \dfrac{g_1(x) - g_2(y)}{f_1(x) - f_2(y)}, \\ v = \dfrac{f_1(x) g_2(y) - f_2(y) g_1(x)}{f_1(x) - f_2(y)}, \end{cases}$$

and we obviously have

(9)
$$\begin{cases} g_1(x) = u f_1(x) + v, \\ g_2(y) = u f_2(y) + v. \end{cases}$$

According to (7) and (9), a homographic transformation (6) of the f_i, g_i transforms u and v by the associated homographic, and this is also true conversely, which is a critical property for what will follow.

We will first obtain a necessary condition for equation (1) to be able to be reduced to the form (4); for this, let us differentiate the first equation in (9) with respect to y, and the second equation with respect to x:

$$\text{(10)} \quad \begin{cases} 0 = \dfrac{\partial u}{\partial y} f_1(x) + \dfrac{\partial v}{\partial y}, \\ 0 = \dfrac{\partial u}{\partial x} f_2(y) + \dfrac{\partial v}{\partial x}. \end{cases}$$

Let us first assume that $\frac{\partial u}{\partial y} = 0$; then, according to (10), $\frac{\partial v}{\partial y} = 0$, that is, $u = u(x)$, $v = v(x)$ and (9) and (7) yield

$$g_2(y) = u(x) f_2(y) + v(x),$$
$$g_3(z) = u(x) f_3(z) + v(x),$$

from which, by letting $x = \text{const.} = x_0$, $u(x_0) = u_0$, $v(x_0) = v_0$,

$$g_2(y) = u_0 f_2(y) + v_0,$$
$$g_3(z) = u_0 f_3(z) + v_0.$$

By subtraction, we obtain

$$[u(x) - u_0] f_2(y) + v(x) - v_0 = 0,$$
$$[u(x) - u_0] f_3(z) + v(x) - v_0 = 0.$$

If $u(x) = u_0$, $v(x) = v_0$, then we also have $g_1(x) = u_0 f_1(x) + v_0$ and (4) can be reduced to an identity; if not, then one would have $f_2(y) = f_3(z) = \text{const.}$, $g_2(y) = g_3(z) = \text{const.}$, and (4) would still be reduced to an identity. The same comment applies to the case of $\frac{\partial u}{\partial x} = 0$.

Leaving aside these trivial cases, (10) yields

$$\text{(11)} \quad \begin{cases} f_1(x) = -\dfrac{\frac{\partial v}{\partial y}}{\frac{\partial u}{\partial y}}, \\ f_2(y) = -\dfrac{\frac{\partial v}{\partial x}}{\frac{\partial u}{\partial x}}. \end{cases}$$

Differentiating the first of these equations with respect to y, and the second with respect to x, we obtain

$$\text{(12)} \quad \begin{cases} \dfrac{\partial^2 v}{\partial y^2} \dfrac{\partial u}{\partial y} - \dfrac{\partial^2 u}{\partial y^2} \dfrac{\partial v}{\partial y} = 0, \\ \dfrac{\partial^2 v}{\partial x^2} \dfrac{\partial u}{\partial x} - \dfrac{\partial^2 u}{\partial x^2} \dfrac{\partial v}{\partial x} = 0. \end{cases}$$

We clearly have
$$\frac{\partial[g_3(z),z]}{\partial(x,y)} = 0;$$
by substituting the expression (7), and observing that
$$\frac{\partial[f_3(z),z]}{\partial(x,y)} = 0,$$
we obtain

(13) $$f_3(x)\frac{\partial(u,z)}{\partial(x,y)} + \frac{\partial(v,z)}{\partial(x,y)} = 0.$$

The hypothesis $\frac{\partial(u,z)}{\partial(x,y)} = 0$ yields $u = u(z)$ and, according to (7), $v = v(z)$; the reasoning that we just employed thus teaches us that (4) is reduced to an identity. Setting this case aside, let us propose that

(14) $$\begin{cases} M = -\dfrac{\frac{\partial z}{\partial y}}{\frac{\partial z}{\partial x}}, \\ N = \dfrac{\partial M}{\partial x} + \dfrac{1}{M}\dfrac{\partial M}{\partial y} = \dfrac{\left(\frac{\partial z}{\partial y}\right)^2 \frac{\partial^2 z}{\partial x^2} - 2\frac{\partial z}{\partial x}\frac{\partial z}{\partial y}\frac{\partial^2 z}{\partial x \partial y} + \left(\frac{\partial z}{\partial x}\right)^2 \frac{\partial^2 z}{\partial y^2}}{\left(\frac{\partial z}{\partial x}\right)^2 \frac{\partial z}{\partial y}}; \end{cases}$$

equation (13) then yields

(15) $$f_3(z) = -\frac{M\frac{\partial v}{\partial x} + \frac{\partial v}{\partial y}}{M\frac{\partial u}{\partial x} + \frac{\partial u}{\partial y}}.$$

Let us assume that $\frac{\partial(u,v)}{\partial(x,y)} = 0$; then (10) yields
$$\frac{\partial u}{\partial x}\frac{\partial u}{\partial y}[f_1(x) - f_2(y)] = 0,$$
and (4) is reduced to an identity. Thus we can posit that

(16) $$\frac{\partial(u,v)}{\partial(x,y)} = \frac{\partial u}{\partial x}\frac{\partial v}{\partial y} - \frac{\partial v}{\partial x}\frac{\partial u}{\partial y} = e^\theta.$$

We will additionally introduce the notations

(17)
$$\begin{cases} A = \left(\dfrac{\partial^2 u}{\partial x^2} \dfrac{\partial v}{\partial x} - \dfrac{\partial^2 v}{\partial x^2} \dfrac{\partial u}{\partial x} \right) e^{-\theta}, \\[6pt] B = \left(\dfrac{\partial^2 u}{\partial y^2} \dfrac{\partial v}{\partial y} - \dfrac{\partial^2 v}{\partial y^2} \dfrac{\partial u}{\partial y} \right) e^{-\theta}, \\[6pt] C = \left(\dfrac{\partial^2 u}{\partial x^2} \dfrac{\partial v}{\partial y} - \dfrac{\partial^2 v}{\partial x^2} \dfrac{\partial u}{\partial y} + 2 \dfrac{\partial^2 u}{\partial x \partial y} \dfrac{\partial v}{\partial x} - 2 \dfrac{\partial^2 v}{\partial x \partial y} \dfrac{\partial u}{\partial x} \right) e^{-\theta}, \\[6pt] D = \left(\dfrac{\partial^2 v}{\partial y^2} \dfrac{\partial u}{\partial x} - \dfrac{\partial^2 u}{\partial y^2} \dfrac{\partial v}{\partial x} + 2 \dfrac{\partial^2 v}{\partial x \partial y} \dfrac{\partial u}{\partial y} - 2 \dfrac{\partial^2 u}{\partial x \partial y} \dfrac{\partial v}{\partial x} \right) e^{-\theta}. \end{cases}$$

The expressions A, B, C, and D are invariant for any homographic transformation of u and v,[3] that is, for any transformation (6) of the f_i and g_i.

Let us once again take up the equation
$$\frac{\partial [f_3(z), z]}{\partial (x, y)} = 0,$$
and in it, substitute expression (15); after all the calculations are done, we obtain
$$M^3 A + B + M^2 C - MD + MN = 0,$$
or, since $A = B = 0$ by virtue of (12),

(18) $$D = MC + N.$$

Differentiating (16) with respect to x and y, we obtain

(19)
$$\begin{cases} \dfrac{\partial^2 u}{\partial x^2} \dfrac{\partial v}{\partial y} - \dfrac{\partial^2 v}{\partial x^2} \dfrac{\partial u}{\partial y} - \left(\dfrac{\partial^2 u}{\partial x \partial y} \dfrac{\partial v}{\partial y} - \dfrac{\partial^2 v}{\partial x \partial y} \dfrac{\partial u}{\partial y} \right) = e^\theta \dfrac{\partial \theta}{\partial x}, \\[6pt] \dfrac{\partial^2 v}{\partial y^2} \dfrac{\partial u}{\partial x} - \dfrac{\partial^2 u}{\partial y^2} \dfrac{\partial v}{\partial x} - \left(\dfrac{\partial^2 v}{\partial x \partial y} \dfrac{\partial u}{\partial y} - \dfrac{\partial^2 u}{\partial x \partial y} \dfrac{\partial v}{\partial y} \right) = e^\theta \dfrac{\partial \theta}{\partial y}, \end{cases}$$

---

[3]E. GOURSAT, *Sur un système d'équations aux dérivées partielles* (*Computes rendus*, v. CIV, 16 mai 1887, p. 1361–1262).– P. PAINLEVÉ, *Sur les équations linéaires simultanées aux dérivées partielles* (*Comptes rendus*, v. CIV, 31 mai 1887, p. 1497–1501).

and these equations, combined with the last two equations of (17), yield

(20)
$$\begin{cases} \dfrac{\partial^2 u}{\partial x^2}\dfrac{\partial v}{\partial y} - \dfrac{\partial^2 v}{\partial x^2}\dfrac{\partial u}{\partial y} = \dfrac{1}{3}\left(C + 2\dfrac{\partial \theta}{\partial x}\right)e^\theta, \\ \dfrac{\partial^2 u}{\partial x \partial y}\dfrac{\partial v}{\partial x} - \dfrac{\partial^2 v}{\partial x \partial y}\dfrac{\partial u}{\partial x} = \dfrac{1}{3}\left(C - \dfrac{\partial \theta}{\partial x}\right)e^\theta, \\ \dfrac{\partial^2 v}{\partial y^2}\dfrac{\partial u}{\partial x} - \dfrac{\partial^2 u}{\partial y^2}\dfrac{\partial v}{\partial x} = \dfrac{1}{3}\left(D + 2\dfrac{\partial \theta}{\partial y}\right)e^\theta, \\ \dfrac{\partial^2 v}{\partial x \partial y}\dfrac{\partial u}{\partial y} - \dfrac{\partial^2 u}{\partial x \partial y}\dfrac{\partial v}{\partial y} = \dfrac{1}{3}\left(D - \dfrac{\partial \theta}{\partial y}\right)e^\theta. \end{cases}$$

Solving equations (12) and (20) with respect to the six second-order derivatives that are involved, we obtain, considering (16),

(21)
$$\begin{cases} \dfrac{\partial^2 u}{\partial x^2} = \dfrac{1}{3}\left(2\dfrac{\partial \theta}{\partial x} + C\right)\dfrac{\partial u}{\partial x}, \\ \dfrac{\partial^2 u}{\partial x \partial y} = \dfrac{1}{3}\left(\dfrac{\partial \theta}{\partial y} - D\right)\dfrac{\partial u}{\partial x} - \dfrac{1}{3}\left(\dfrac{\partial \theta}{\partial x} - C\right)\dfrac{\partial u}{\partial y}, \\ \dfrac{\partial^2 u}{\partial y^2} = \dfrac{1}{3}\left(2\dfrac{\partial \theta}{\partial y} + D\right)\dfrac{\partial u}{\partial y}, \end{cases}$$

and precisely the same system of equations for v.

Let us set down the conditions for the integratability of this system, namely
$$\dfrac{\partial}{\partial y}\dfrac{\partial^2 u}{\partial x^2} = \dfrac{\partial}{\partial x}\dfrac{\partial^2 u}{\partial x \partial y}, \quad \dfrac{\partial}{\partial y}\dfrac{\partial^2 u}{\partial x \partial y} = \dfrac{\partial}{\partial x}\dfrac{\partial^2 u}{\partial y^2};$$
replacing the second order derivatives by the values taken from (21), we obtain

$$\left[3\dfrac{\partial^2 \theta}{\partial x \partial y} + 3\dfrac{\partial C}{\partial y} + 3\dfrac{\partial D}{\partial x} - \left(\dfrac{\partial \theta}{\partial x} - C\right)\left(\dfrac{\partial \theta}{\partial y} - D\right)\right]\dfrac{\partial u}{\partial x}$$
$$- \left[3\dfrac{\partial^2 \theta}{\partial x^2} - 3\dfrac{\partial C}{\partial x} - \left(\dfrac{\partial \theta}{\partial x} - C\right)\left(\dfrac{\partial \theta}{\partial x} + 2C\right)\right]\dfrac{\partial u}{\partial y} = 0,$$
$$\left[3\dfrac{\partial^2 \theta}{\partial y^2} - 3\dfrac{\partial D}{\partial y} - \left(\dfrac{\partial \theta}{\partial y} - D\right)\left(\dfrac{\partial \theta}{\partial y} + 2D\right)\right]\dfrac{\partial u}{\partial x}$$
$$- \left[3\dfrac{\partial^2 \theta}{\partial x \partial y} + 3\dfrac{\partial C}{\partial y} + 3\dfrac{\partial D}{\partial x} - \left(\dfrac{\partial \theta}{\partial x} - C\right)\left(\dfrac{\partial \theta}{\partial y} - D\right)\right]\dfrac{\partial u}{\partial y} = 0.$$

The same equations being satisfied by v, and the functional determinant of u and v being different from zero, it follows that the coefficients of $\dfrac{\partial u}{\partial x}$ and $\dfrac{\partial u}{\partial y}$

cancel out in the preceding equations, such that

$$
(22) \quad \begin{cases} \dfrac{\partial^2 \theta}{\partial x^2} = \dfrac{1}{3}\left(\dfrac{\partial \theta}{\partial x} - C\right)\left(\dfrac{\partial \theta}{\partial x} + 2C\right) + \dfrac{\partial C}{\partial x}, \\ \dfrac{\partial^2 \theta}{\partial x \partial y} = \dfrac{1}{3}\left(\dfrac{\partial \theta}{\partial x} - C\right)\left(\dfrac{\partial \theta}{\partial y} - D\right) - \dfrac{\partial C}{\partial y} - \dfrac{\partial D}{\partial x}, \\ \dfrac{\partial^2 \theta}{\partial y^2} = \dfrac{1}{3}\left(\dfrac{\partial \theta}{\partial y} - D\right)\left(\dfrac{\partial \theta}{\partial y} + 2D\right) + \dfrac{\partial D}{\partial y}. \end{cases}
$$

Let us set down the conditions for the integratability of this system, namely

$$\frac{\partial}{\partial y}\frac{\partial^2 \theta}{\partial x^2} = \frac{\partial}{\partial x}\frac{\partial^2 \theta}{\partial x \partial y}, \quad \frac{\partial}{\partial y}\frac{\partial^2 \theta}{\partial x \partial y} = \frac{\partial}{\partial x}\frac{\partial^2 \theta}{\partial y^2};$$

by replacing the second derivatives by their values, we can see that θ disappears entirely from these equations, which turn into

$$
(23) \quad \begin{cases} 2\dfrac{\partial^2 C}{\partial x \partial y} + \dfrac{\partial^2 D}{\partial x^2} - C\left(2\dfrac{\partial C}{\partial y} + \dfrac{\partial D}{\partial x}\right) = 0, \\ \dfrac{\partial^2 C}{\partial y^2} + 2\dfrac{\partial^2 D}{\partial x \partial y} - D\left(\dfrac{\partial C}{\partial y} + 2\dfrac{\partial D}{\partial x}\right) = 0. \end{cases}
$$

Substituting the value of D given by (18), we obtain[4]

$$
(24) \quad \begin{cases} M\dfrac{\partial^2 C}{\partial x^2} + 2\dfrac{\partial^2 C}{\partial x \partial y} = \left(MC - 2\dfrac{\partial M}{\partial x}\right)\dfrac{\partial C}{\partial x} + 2C\dfrac{\partial C}{\partial y} \\ \qquad + \dfrac{\partial M}{\partial x}C^2 + \left(\dfrac{\partial N}{\partial x} - \dfrac{\partial^2 M}{\partial x^2}\right)C - \dfrac{\partial^2 N}{\partial x^2}, \\ 2M\dfrac{\partial^2 C}{\partial x \partial y} + \dfrac{\partial^2 C}{\partial y^2} = 2\left(M^2 C + MN - \dfrac{\partial M}{\partial y}\right)\dfrac{\partial C}{\partial x} \\ \qquad + \left(MC + N - 2\dfrac{\partial M}{\partial x}\right)\dfrac{\partial C}{\partial y} \\ \qquad + 2M\dfrac{\partial M}{\partial x}C^2 + 2\left(N\dfrac{\partial M}{\partial x} + M\dfrac{\partial N}{\partial x} - \dfrac{\partial^2 M}{\partial x \partial y}\right)C \\ \qquad + 2N\dfrac{\partial N}{\partial x} - 2\dfrac{\partial^2 N}{\partial x \partial y}. \end{cases}
$$

If, as a consequence, the given equation (1) is reducible to form (4), the two equations have a common integral, which we can clearly assume to be analytic. I say this condition is also sufficient.

---

[4]Two corrections dictated by the derivation have been applied to the second equation in (24): a factor of 2 on $\frac{\partial M}{\partial x}$ in the second term and a factor of 2 on the last term. – EDITOR

2. ON EQUATIONS OF THREE VARIABLES

To demonstrate this, let us begin by transforming equations (21) and (22) into a single linear system. To accomplish this, let us posit in (21) that

(25) $$u = e^{\frac{1}{3}\theta}\omega.$$

As a result, considering (22),

(26) $$\begin{cases} \dfrac{\partial^2\omega}{\partial x^2} = \dfrac{1}{3}C\dfrac{\partial\omega}{\partial x} + \dfrac{1}{3}\left(\dfrac{2}{3}C^2 - \dfrac{\partial C}{\partial x}\right)\omega, \\ \dfrac{\partial^2\omega}{\partial x\partial y} = -\dfrac{1}{3}D\dfrac{\partial\omega}{\partial x} - \dfrac{1}{3}C\dfrac{\partial\omega}{\partial y} + \dfrac{1}{3}\left(-\dfrac{1}{3}CD + \dfrac{\partial C}{\partial y} + \dfrac{\partial D}{\partial x}\right)\omega, \\ \dfrac{\partial^2\omega}{\partial y^2} = \dfrac{1}{3}D\dfrac{\partial\omega}{\partial y} + \dfrac{1}{3}\left(\dfrac{2}{3}D^2 - \dfrac{\partial D}{\partial y}\right)\omega. \end{cases}$$

Furthermore, the substitution

(27) $$e^{-\frac{1}{3}\theta} = \omega,$$

transforms the system (22) into the same system (26). The conditions for the integrability of (22) and (26) are thus the same, namely the equations (23).

Let us assume that these equations have been verified; since (26) allows us to express any derivative of ω as a linear function of $\frac{\partial\omega}{\partial x}$, $\frac{\partial\omega}{\partial y}$ and ω, it follows, according to well-known theories, that any integral of (26) can be expressed linearly with constant coefficients in the three particular integrals, $\omega_1, \omega_2, \omega_3$,

$$\omega = \alpha\omega_1 + \beta\omega_2 + \gamma\omega_3,$$

the fundamental system $\omega_1, \omega_2, \omega_3$ being such that the determinant

(28) $$\Delta(\omega_1, \omega_2, \omega_3) = \begin{vmatrix} \omega_1 & \frac{\partial\omega_1}{\partial x} & \frac{\partial\omega_1}{\partial y} \\ \omega_2 & \frac{\partial\omega_2}{\partial x} & \frac{\partial\omega_2}{\partial y} \\ \omega_3 & \frac{\partial\omega_3}{\partial x} & \frac{\partial\omega_3}{\partial y} \end{vmatrix} = \omega_3^3\frac{\partial(\frac{\omega_1}{\omega_3}, \frac{\omega_2}{\omega_3})}{\partial(x, y)}$$

will not cancel out identically. From (28) and (26), it follows that

$$\frac{\partial\Delta}{\partial x} = \frac{1}{3}C\Delta - \frac{1}{3}C\Delta = 0,$$
$$\frac{\partial\Delta}{\partial y} = -\frac{1}{3}D\Delta + \frac{1}{3}D\Delta = 0;$$

we thus have

(29) $$\Delta(\omega_1, \omega_2, \omega_3) = \text{const.} = \frac{1}{k^3} \neq 0.$$

Let us set

(30) $$u = \frac{\omega_1}{\omega_3}, \quad v = \frac{\omega_2}{\omega_3}, \quad \theta = -3\log(k\omega_3);$$

according to (25), (27), (28), and (29), u, v, and θ satisfy equations (16), (21 for u), (21 for v) and (22).

Let $\bar{\omega}_1$, $\bar{\omega}_2$, $\bar{\omega}_3$ be another fundamental system of (26); we then have

(31) $$\bar{\omega}_i = \alpha_i \omega_1 + \beta_i \omega_2 + \gamma_i \omega_3 \quad (i = 1, 2, 3),$$

where

$$\begin{vmatrix} \alpha_1 & \beta_1 & \gamma_1 \\ \alpha_2 & \beta_2 & \gamma_2 \\ \alpha_3 & \beta_3 & \gamma_3 \end{vmatrix} \neq 0$$

and

(32) $$\Delta(\bar{\omega}_1, \bar{\omega}_2, \bar{\omega}_3) = \begin{vmatrix} \alpha_1 & \beta_1 & \gamma_1 \\ \alpha_2 & \beta_2 & \gamma_2 \\ \alpha_3 & \beta_3 & \gamma_3 \end{vmatrix} \Delta(\omega_1, \omega_2, \omega_3) = \frac{1}{\bar{k}^3} \neq 0.$$

Setting

(33) $$\bar{u} = \frac{\bar{\omega}_1}{\bar{\omega}_3}, \quad \bar{v} = \frac{\bar{\omega}_2}{\bar{\omega}_3}, \quad \bar{\theta} = -3 \log(\bar{k} \bar{\omega}_3),$$

$\bar{u}$, $\bar{v}$, and $\bar{\theta}$ satisfy the same equations as just before and, by virtue of (31),

(34) $$\begin{cases} \bar{u} = \dfrac{\alpha_1 u + \beta_1 v + \gamma_1}{\alpha_3 u + \beta_3 v + \gamma_3}, \\ \bar{v} = \dfrac{\alpha_2 u + \beta_2 v + \gamma_2}{\alpha_3 u + \beta_3 v + \gamma_3}. \end{cases}$$

Let us conversely set u, v and θ as functions that satisfy (16), (21 for u), (21 for v), and (22); in setting

(35) $$\omega_1 = u e^{-\frac{1}{3}\theta}, \quad \omega_2 = v e^{-\frac{1}{3}\theta}, \quad \omega_3 = e^{-\frac{1}{3}\theta},$$

these expressions satisfy (26) and constitute a fundamental system because (28) yields

$$\Delta(\omega_1, \omega_2, \omega_3) = 1.$$

Any solution $\bar{u}$, $\bar{v}$, and $\bar{\theta}$ of the equations for u, v and θ gives rise, according to (35), to another fundamental system, $\bar{\omega}_1$, $\bar{\omega}_2$, $\bar{\omega}_3$; this system being connected to the first system by relations of the form (31), and we can see that $\bar{u}$ and $\bar{v}$ can be deduced from u and v by homography (34).

These points having been established, our demonstration concludes in the following way. If the two equations (24) have a common integral C, the equations (23) will allow for a pair C, D connected by the relation (18). There thus exists a fundamental system $\omega_1, \omega_2, \omega_3$ of (26), and the formulas (30) determine a u and a v that satisfy (16), (21 for u), (21 for v), (22), as well as equations (19) derived from (16). In addition, u and v satisfy (12) which are the linear combinations of the preceding equations. Let us now determine two functions, f_1 and f_2, of the formulas (11); according to (12), f_1 will be a function only of x and f_2 only of y.

Next, let us calculate two functions, g_1 and g_2, using (9); according to (11), they will depend solely on x and y, respectively. Equation (15) yields an f_3 which, as a consequence of (18), is solely a function of z; finally (7) gives us a g_3, according to either (15) or (13), dependent only upon z. Since (7) is identical to (4), we have thus reduced the given equation (1) to the desired form.

Beginning from another fundamental system, $\bar{\omega}_1, \bar{\omega}_2, \bar{\omega}_3$ we will have found two other functions, $\bar{u}$ and $\bar{v}$, and, by degrees, obtained the other functions, $\bar{f}_i, \bar{g}_i (i = 1, 2, 3)$ in the manner indicated. But, according to (34), $\bar{u}$ and $\bar{v}$ are homographic transformations of u and of v, and as a consequence, the $\bar{f}_i, \bar{g}_i$ are transformations of f_i, g_i associated with (34) by homography (6).

In summary, we have proven the following fundamental theorem:

The necessary and sufficient condition in order that the given equation (1) may be reduced into the form (4) consists of the existence of a common integral C for the two partial differential equations (24).

All equations (4) belonging to the same value of C can be derived from any one of their number by means of a homography (6), and conversely, two equations (4) that are homographic lead to the same value of C.

2. – Conditions for One or More Scales to be Rectilinear

Graphical plotting of a nomogram is simplified to a considerable degree when one or several scales become rectilinear, a circumstance that one encounters in many of the equations that arise in practice. We will first investigate the necessary and sufficient condition for the scale of x to be rectilinear.

By an appropriate homographic transformation, it is possible to reduce the line supporting the scale of x to become the axis of η; in equation (4), we thus have $f_1(x) = 0$. The first of the equations (9) thus yields

$$\frac{\partial v}{\partial y} = 0.$$

Since $\frac{\partial v}{\partial x} \neq 0$ as a consequence of (16), the second of the equations (21) yields

(36) $$\frac{\partial \theta}{\partial y} = D,$$

and the second equation (22)

(37) $$\frac{\partial C}{\partial y} + 2\frac{\partial D}{\partial x} = 0.$$

Equation (37), combined with equations (18) and (23), thus constitutes a *necessary condition for the scale of x to be rectilinear*. I state in addition that this condition is also sufficient.

First, equations (23) and (37) are equivalent to the system

(38) $$\begin{cases} \dfrac{\partial C}{\partial y} + 2\dfrac{\partial D}{\partial x} = 0, \\ \dfrac{\partial^2 C}{\partial x \partial y} = C\dfrac{\partial C}{\partial y}. \end{cases}$$

If x_0, y_0 is a point where C and D are holomorphic, the second of these equations yields, by integrating in relation to y between the limits y_0 and y, and setting $C(x, y_0) = C_0$,

(39) $$\frac{\partial C}{\partial x} - \frac{\partial C_0}{\partial x} = \frac{1}{2}\left(C^2 - C_0^2\right).$$

Taking equation (36) into account, equations (22) may be reduced to

$$\frac{\partial^2 \theta}{\partial x^2} = \frac{1}{3}\left(\frac{\partial \theta}{\partial x} - C\right)\left(\frac{\partial \theta}{\partial x} + 2C\right) + \frac{\partial C}{\partial x},$$

$$\frac{\partial \theta}{\partial y} = D;$$

the expression

(40) $$\theta = \int_{x_0, y_0}^{x, y} \left(\frac{3}{2}C_0 - \frac{1}{2}C\right) dx + D\, dy$$

constitutes a particular integral of the preceding system, because, first of all, the expression under the integral sign is a precise differential by virtue of (37), and thus θ, which clearly satisfies the terms of the second equation of the system, also satisfies the first on account of (39).

Substituting this value of θ in system (21), this becomes

(41) $$\begin{cases} \dfrac{\partial^2 u}{\partial x^2} = C_0 \dfrac{\partial u}{\partial x}, \\ \dfrac{\partial^2 u}{\partial x \partial y} = \dfrac{1}{2}(C_0 - C)\dfrac{\partial u}{\partial y}, \\ \dfrac{\partial^2 u}{\partial y^2} = D\dfrac{\partial u}{\partial y}, \end{cases}$$

and we must find a u and a v each of which satisfy this system, satisfy (16) and finally satisfy $\frac{\partial v}{\partial y} = 0$. Let us take as v any integral of the system

$$\frac{\partial v}{\partial x} = e^{\int_{x_0}^{x} C_0 dx},$$

$$\frac{\partial v}{\partial y} = 0;$$

equation (16) thus yields

(42) $$\frac{\partial u}{\partial y} = -e^{\int_{x_0,y_0}^{x,y} \frac{1}{2}(C_0-C)dx+Ddy},$$

an expression that satisfies the two last equations (41), of which the first equation additionally yields

(43) $$\frac{\partial u}{\partial x} = \varphi(y) e^{\int_{x_0}^{x} C_0 dx}.$$

The condition of integrability of (42) and (43), namely

$$\frac{\partial^2 u}{\partial x \partial y} = \frac{\partial^2 u}{\partial y \partial x},$$

yields for determining φ, the equation

(44) $$\frac{\partial \varphi}{\partial y} = -\frac{1}{2}(C_0 - C) e^{\int_{x_0,y_0}^{x,y} -\frac{1}{2}(C_0+C)dx+Ddy},$$

whose right term is solely a function of y, as we can see by differentiating according to x and using (39). Setting u as any particular integral of (42) and (43), we thus have three functions, u, v, and θ that satisfy (16), (21 for u), (21 for v), (22), and, in addition, $\frac{\partial v}{\partial y} = 0$. Subsequently determining the f_i, g_i, $(i = 1, 2, 3)$ by formulas (11), (9), (15), and (7), we can see that $f_1(x) = 0$; therefore, the scale of x is rectilinear.

Permutating x and y, the formulas of Section 1 show that

$$u, \quad v, \quad \theta, \quad C, \quad D, \quad M$$

are transformed into

$$u, \quad v, \quad -\theta, \quad D, \quad C, \quad \frac{1}{M},$$

and from this, it follows that *the necessary and sufficient condition for the scale of y to be rectilinear is expressed by the equation*

(45) $$2\frac{\partial C}{\partial y} + \frac{\partial D}{\partial x} = 0,$$

combined with equations (18) *and* (23).

Finally, to determine the condition for which the scale of z is rectilinear, one simply needs to take y and z as independent variables. By setting, as is customary,

$$p = \frac{\partial z}{\partial x}, \quad q = \frac{\partial z}{\partial y}, \quad r = \frac{\partial^2 z}{\partial x^2}, \quad s = \frac{\partial^2 z}{\partial x \partial y}, \quad t = \frac{\partial^2 z}{\partial y^2},$$

and denoting the differentiations with respect to the new independent variables as $\frac{\delta}{\delta y}$, $\frac{\delta}{\delta z}$ as well as by M_x, N_x, θ_x, C_x, D_x, the expressions that replace M, N, θ, C, D, namely

$$M_x = -\frac{\frac{\delta x}{\delta z}}{\frac{\delta x}{\delta y}}, \quad N_x = \frac{\delta M_x}{\delta y} + \frac{1}{M_x}\frac{\delta M_x}{\delta z},$$

$$e^{\theta_x} = \frac{\delta u}{\delta y}\frac{\delta v}{\delta z} - \frac{\delta u}{\delta z}\frac{\delta v}{\delta y},$$

$$C_x = \left(\frac{\delta^2 u}{\delta y^2}\frac{\delta v}{\delta z} - \frac{\delta^2 v}{\delta y^2}\frac{\delta u}{\delta z} + 2\frac{\delta^2 u}{\delta y \delta z}\frac{\delta v}{\delta y} - 2\frac{\delta^2 v}{\delta y \delta z}\frac{\delta u}{\delta y}\right)e^{-\theta_x},$$

$$D_x = \left(\frac{\delta^2 v}{\delta z^2}\frac{\delta u}{\delta y} - \frac{\delta^2 u}{\delta z^2}\frac{\delta v}{\delta y} + 2\frac{\delta^2 v}{\delta y \delta z}\frac{\delta u}{\delta z} - 2\frac{\delta^2 v}{\delta y \delta z}\frac{\delta u}{\delta z}\right)e^{-\theta_x},$$

the well known elementary formulas

$$\frac{\delta}{\delta y} = \frac{\partial}{\partial y} - \frac{q}{p}\frac{\partial}{\partial x},$$

$$\frac{\delta}{\delta z} = \frac{1}{p}\frac{\partial}{\partial x}$$

yield, taking (12) and (18) into account,

(46)
$$\begin{cases} M_x = \frac{1}{q}, \\ N_x = -\frac{t}{q^2}, \\ e^{\theta_x} = -\frac{1}{p}e^{\theta}, \\ C_x = \frac{q}{p}C - \frac{qr}{p^2} + \frac{t}{q} = -D - \frac{2s}{p} + \frac{2t}{q}, \\ D_x = \frac{1}{p}C - \frac{r}{p^2} = -\frac{1}{q}D - \frac{2s}{pq} + \frac{t}{q^2}, \\ D_x = M_x C_x + N_x, \\ 2\frac{\delta C_x}{\delta z} + \frac{\delta D_x}{\delta y} = \frac{1}{p}\left(\frac{\partial C}{\partial y} - \frac{\partial D}{\partial x}\right) + \frac{3}{p}\frac{\partial^2 \log M}{\partial x \partial y}, \\ \frac{\delta C_x}{\delta z} + 2\frac{\delta D_x}{\delta y} = \frac{1}{p}\left(2\frac{\partial C}{\partial y} + \frac{\partial D}{\partial x}\right). \end{cases}$$

2. ON EQUATIONS OF THREE VARIABLES

Moreover, it is clear that equations (23), by their signification as necessary and sufficient conditions [with (18) which, according to (46) is transformed into itself] for (1) to be reducible to the form (4), remain invariable for the changes in variables that are at issue. It follows, according to (45) and (46), that *the necessary and sufficient condition* for the scale of z to be rectilinear is expressed by the equation

$$(47) \qquad \frac{\partial C}{\partial y} - \frac{\partial D}{\partial x} + 3\frac{\partial^2 \log M}{\partial x \partial y} = 0,$$

combined, of course, *with equations* (18) *and* (23).

According to what we have just seen, the conditions for several of the scales to be rectilinear can be obtained by combining the conditions for individual scales to be rectilinear. For example, equations (37) and (45) yield

$$(48) \qquad \frac{\partial C}{\partial y} = \frac{\partial D}{\partial x} = 0,$$

and equations (23) are thus satisfied in identical fashion; it follows that (48) *and* (18) *express the necessary and sufficient conditions for both scales x and y to be simultaneously rectilinear.*

Finally, equations (37), (45) and (47) yield

$$(49) \qquad \frac{\partial C}{\partial y} = \frac{\partial D}{\partial x} = \frac{\partial^2 \log M}{\partial x \partial y} = 0.$$

The equation

$$(50) \qquad \frac{\partial^2 \log M}{\partial x \partial y} = 0$$

expresses the necessary and sufficient condition so that the given equation (1) *permits nomographic representation with three rectilinear scales.*

We can directly demonstrate that the equations $\frac{\partial C}{\partial y} = \frac{\partial D}{\partial x} = 0$, $D = MC + N$ possess a common integral when $\frac{\partial^2 \log M}{\partial x \partial y} = 0$; however, it is more worthwhile, given the developments that will follow in Section 4, to proceed in the following manner. The integral of (50) is clearly

$$M = -\frac{\psi'(y)}{\varphi'(x)},$$

$\varphi(x)$ and $\psi(x)$ being arbitrary functions. The first of equations (14) thus become

$$\frac{\partial[z, \varphi(x) + \psi(y)]}{\partial(x, y)} = 0,$$

whose general integral is obviously
(51) $$\varphi(x) + \psi(y) + \chi(z) = 0,$$
$\chi(z)$ designating a new arbitrary function. Now, (51) may be written as[5]

$$\begin{vmatrix} \varphi(x) & -1 & 1 \\ \psi(y) & 1 & 1 \\ -\frac{1}{2}\chi(z) & 0 & 1 \end{vmatrix} = 0,$$

such that condition (50), necessary according to (49), is also sufficient.

3. – About the Determination of u, v, and θ when C is Known

Let us assume that the scales x and y are both curved, that is to say, that

(52) $$\frac{\partial C}{\partial y} + 2\frac{\partial D}{\partial x} \neq 0, \quad 2\frac{\partial C}{\partial y} + \frac{\partial D}{\partial x} \neq 0.$$

Next, we will see how, assuming that C is known, u, v, and θ, and thus f_i and g_i can be derived by differentiations and eliminations. The question obviously comes back to the examination of a fundamental system ω_1, ω_2, ω_3, of (26).

The first equation (26) can be written as

$$\frac{\partial}{\partial x}\left(\frac{\partial \omega}{\partial x} + \frac{1}{3}C\omega\right) = \frac{2}{3}C\left(\frac{\partial \omega}{\partial x} + \frac{1}{3}C\omega\right),$$

from which,

$$\frac{\partial \omega}{\partial x} + \frac{1}{3}C\omega = \varphi(y)e^{\frac{2}{3}\int C\,dx}.$$

Now, according to (52), the first of equations (23) can be written as

(53) $$C = \frac{\partial}{\partial x}\log\left(2\frac{\partial C}{\partial y} + \frac{\partial D}{\partial x}\right),$$

such that we obtain

(54) $$\frac{\partial \omega}{\partial x} + \frac{1}{3}C\omega = \varphi(y)\left(2\frac{\partial C}{\partial y} + \frac{\partial D}{\partial x}\right)^{\frac{2}{3}}.$$

The general integral of (54) is

$$\omega = e^{-\frac{1}{3}\int C\,dx}\left[\psi(y) + \varphi(y)\int e^{\frac{1}{3}\int C\,dx}\left(2\frac{\partial C}{\partial y} + \frac{\partial D}{\partial x}\right)^{\frac{2}{3}}dx\right],$$

---

[5]The first entry in the second row was given by Gronwall as $\psi(x)$ but the mathematics dictate that the correct entry is $\psi(y)$. – EDITOR

or, according to (53),
$$\omega = \frac{1}{\left(2\frac{\partial C}{\partial y} + \frac{\partial D}{\partial x}\right)^{\frac{1}{3}}} \left[\psi(y) + \varphi(y) \int \left(2\frac{\partial C}{\partial y} + \frac{\partial D}{\partial x}\right) dx\right],$$

or, applying (53) once more,

(55) $$\omega = \frac{1}{\left(2\frac{\partial C}{\partial y} + \frac{\partial D}{\partial x}\right)^{\frac{1}{3}}} \left\{\psi(y) + \varphi(y) \left[2\frac{\partial}{\partial y}\log\left(2\frac{\partial C}{\partial y} + \frac{\partial D}{\partial x}\right) + D\right]\right\}.$$

The second equation (26) may also be written as
$$\frac{\partial}{\partial y}\left(\frac{\partial \omega}{\partial x} + \frac{1}{3}C\omega\right) + \frac{1}{3}D\left(\frac{\partial \omega}{\partial x} + \frac{1}{3}C\omega\right) = \frac{1}{3}\left(2\frac{\partial C}{\partial y} + \frac{\partial D}{\partial x}\right)\omega,$$

and by substituting (54) and (55), after some reductions, we obtain
$$\psi(y) = 3\varphi'(y).$$

From (55), we subsequently obtain

(56) $$\omega = \frac{1}{\left(2\frac{\partial C}{\partial y} + \frac{\partial D}{\partial x}\right)^{\frac{1}{3}}} \left\{3\varphi'(y) + \varphi(y)\left[2\frac{\partial}{\partial y}\log\left(2\frac{\partial C}{\partial y} + \frac{\partial D}{\partial x}\right) + D\right]\right\}.$$

By substituting this expression in the last equation (26), we then obtain a linear and homogeneous third order differential equation for $\varphi(y)$:
$$H(\varphi; x, y) = 0,$$
whose coefficients are dependent, a priori, on x and y. Now, there are three linearly independent functions satisfying $H = 0$, $\varphi_1(y)$, $\varphi_2(y)$ and $\varphi_3(y)$, which correspond to a fundamental system, $\omega_1, \omega_2, \omega_3$, of (26); it follows that, in this last equation, the relations between the coefficients are independent of x. Thus we obtain an equation $H(\varphi; x_i, y) = 0$ equivalent to $H(\varphi; x, y) = 0$, by setting $x = x_i = $ const. in the last equation (26) and substituting the expression (56) for ω, after having set $x = x_i$. But by setting $\omega(x_i, y) = 0$, that is, according to (56),

(57) $$3\varphi'(y) + \varphi(y)\left[2\frac{\partial}{\partial y}\log\left(2\frac{\partial C}{\partial y} + \frac{\partial D}{\partial x}\right) + D\right]_{x=x_i} = 0,$$

the last equation (26) with $x = x_i$ is satisfied, and as a consequence, the equation $H(\varphi; x_i, y) = 0$ or its equivalent $H(\varphi; x, y) = 0$ as well. Since the second equation (23) may be written as

(58) $$D = \frac{\partial}{\partial y}\log\left(\frac{\partial C}{\partial y} + 2\frac{\partial D}{\partial x}\right),$$

it is apparent that a particular integral of (57) is given by the expression

$$(59) \qquad \varphi_i(y) = \cfrac{1}{\left(2\frac{\partial C}{\partial y} + \frac{\partial D}{\partial x}\right)^{\frac{2}{3}}_{x=x_i} \left(\frac{\partial C}{\partial y} + 2\frac{\partial D}{\partial x}\right)^{\frac{1}{3}}_{x=x_i}},$$

and, by virtue of (56), (57), and (59), the expressions

$$(60) \qquad \begin{aligned} \omega_i &= \cfrac{1}{\left(2\frac{\partial C}{\partial y} + \frac{\partial D}{\partial x}\right)^{\frac{2}{3}}_{x=x_i} \left(\frac{\partial C}{\partial y} + 2\frac{\partial D}{\partial x}\right)^{\frac{1}{3}}_{x=x_i} \left(2\frac{\partial C}{\partial y} + \frac{\partial D}{\partial x}\right)^{\frac{1}{3}}} \\ &\times \left\{ 2\frac{\partial}{\partial y}\log\left(2\frac{\partial C}{\partial y} + \frac{\partial D}{\partial x}\right) + D \right. \\ &\quad \left. - \left[2\frac{\partial}{\partial y}\log\left(2\frac{\partial C}{\partial y} + \frac{\partial D}{\partial x}\right) + D\right]_{x=x_i} \right\}, \qquad (i=1,2,3) \end{aligned}$$

are particular integrals of system (26).

Is it possible for us to determine the constants x_1, x_2, x_3 in (60) such that $\omega_1, \omega_2, \omega_3$ constitute a fundamental system? In order to do this, it is necessary and sufficient that $\Delta(\omega_1, \omega_2, \omega_3) \neq 0$. Referring back to (28), it is clear that

$$\Delta(\omega_1, \omega_2, \omega_3) = \frac{1}{\rho^3} \Delta(\rho\omega_1, \rho\omega_2, \rho\omega_3),$$

for any function $\rho = \left(2\frac{\partial C}{\partial y} + \frac{\partial D}{\partial x}\right)^{\frac{1}{3}}$, and (56) gives us

$$\Delta(\omega_1, \omega_2, \omega_3) = \cfrac{1}{2\frac{\partial C}{\partial y} + \frac{\partial D}{\partial x}} \begin{vmatrix} 3\varphi_i'(y) + \varphi_i(y)\left[2\frac{\partial}{\partial y}\log\left(2\frac{\partial C}{\partial y} + \frac{\partial D}{\partial x}\right) + D\right], \\ \varphi_i(y)\left(2\frac{\partial C}{\partial y} + \frac{\partial D}{\partial x}\right), \\ 3\varphi_i''(y) + \varphi_i'(y)\left[2\frac{\partial}{\partial y}\log\left(2\frac{\partial C}{\partial y} + \frac{\partial D}{\partial x}\right) + D\right] \\ + \varphi_i(y)\left[2\frac{\partial^2}{\partial y^2}\log\left(2\frac{\partial C}{\partial y} + \frac{\partial D}{\partial x}\right) + \frac{\partial D}{\partial y}\right] \end{vmatrix}_{(i=1,2,3)}$$

or, by reducing the determinant,

$$\Delta(\omega_1, \omega_2, \omega_3) = -9 \begin{vmatrix} \varphi_1(y) & \varphi_2(y) & \varphi_3(y) \\ \varphi_1'(y) & \varphi_2'(y) & \varphi_3'(y) \\ \varphi_1''(y) & \varphi_2''(y) & \varphi_3''(y) \end{vmatrix}.$$

By setting $\varphi_i(y) = e^{\psi_i(y)}$, we find that

(61) $$\Delta(\omega_1, \omega_2, \omega_3) = -9e^{\psi_1(y)+\psi_2(y)+\psi_3(y)} \begin{vmatrix} 1 \\ \psi_i'(y) \\ \psi_i''(y) + [\psi_i'(y)]^2 \end{vmatrix}_{i=1,2,3}$$

and, according to (57),

$$\psi_i'(y) = -\frac{2}{3}\frac{\partial}{\partial y}\log\left(2\frac{\partial C}{\partial y} + \frac{\partial D}{\partial x}\right)_{x=x_i} - \frac{1}{3}D(x_i, y).$$

Let us first select x_1 and x_2, such that

(62) $$\begin{vmatrix} 1 & 1 \\ \psi_1'(y) & \psi_2'(y) \end{vmatrix} = \psi_2'(y) - \psi_1'(y) \neq 0,$$

something that is always possible, since otherwise, $2\frac{\partial}{\partial y}\log\left(2\frac{\partial C}{\partial y} + \frac{\partial D}{\partial x}\right) + D$ would be independent of x, which is to say

$$\frac{\partial}{\partial x}\left[2\frac{\partial}{\partial y}\log\left(2\frac{\partial C}{\partial y} + \frac{\partial D}{\partial x}\right) + D\right] = 2\frac{\partial C}{\partial y} + \frac{\partial D}{\partial x} = 0,$$

which is contrary to hypothesis (52). With these values of x_1 and x_2, let us assume that expression (61) is canceled out regardless of the value of x_3; by writing x in place of x_3, and $\psi_3(x, y)$ in place of $\psi_3(y)$, we would thus have

$$\begin{vmatrix} 1 & 1 & 1 \\ \psi_1'(y) & \psi_2'(y) & \frac{\partial \psi(x,y)}{\partial y} \\ \psi_1''(y) + \psi_1'(y)^2 & \psi_2''(y) + \psi_2'(y)^2 & \frac{\partial^2 \psi(x,y)}{\partial y^2} + \left[\frac{\partial \psi(x,y)}{\partial y}\right]^2 \end{vmatrix} = 0,$$

or, by developing and in consideration of (62),

(63) $$\frac{\partial^2 \psi(x,y)}{\partial y^2} + \left[\frac{\partial \psi(x,y)}{\partial y}\right]^2 - \alpha(y)\frac{\partial \psi(x,y)}{\partial y} + \beta(y).$$

Let us differentiate this equation with respect to x; since we have

$$\frac{\partial \psi(x,y)}{\partial y} = -\frac{1}{3}\left[2\frac{\partial}{\partial y}\log\left(2\frac{\partial C}{\partial y} + \frac{\partial D}{\partial x}\right) + D\right],$$

$$\frac{\partial}{\partial x}\frac{\partial \psi(x,y)}{\partial y} = -\frac{1}{3}\left(2\frac{\partial C}{\partial y} + \frac{\partial D}{\partial x}\right),$$

it follows that

(64) $$2\frac{\partial^2 C}{\partial y^2} + \frac{\partial^2 C}{\partial x \partial y} + 2D\left(2\frac{\partial C}{\partial y} + \frac{\partial D}{\partial x}\right) = -3\alpha(y)\left(2\frac{\partial C}{\partial y} + \frac{\partial D}{\partial x}\right).$$

Let us differentiate again with respect to x; in consideration of (23), we obtain

$$\frac{\partial}{\partial y}\left[C\left(2\frac{\partial C}{\partial y}+\frac{\partial D}{\partial x}\right)\right]+2\frac{\partial D}{\partial x}\left(2\frac{\partial C}{\partial y}+\frac{\partial D}{\partial x}\right)+2DC\left(2\frac{\partial C}{\partial y}+\frac{\partial D}{\partial x}\right)$$
$$=-3\alpha(y)C\left(2\frac{\partial C}{\partial y}+\frac{\partial D}{\partial x}\right),$$

and eliminating $\alpha(y)$ with the help of (64), we finally come to

$$\left(\frac{\partial C}{\partial y}+2\frac{\partial D}{\partial x}\right)\left(2\frac{\partial C}{\partial y}+\frac{\partial D}{\partial x}\right)=0,$$

contrary to hypothesis (52).

As a consequence, if scales x and y are curved, we can select constants x_1, x_2, x_3 such that in calculating w_1, w_2, w_3 by formula (60), these functions form a fundamental system of (26); the formulas in Section 1 thus yield f_i, g_i by differentiation and elimination.

If, for example, the x scale is rectilinear, and the y and z scales are curved, we would take y and z as independent variables; the method that we just used is thus applicable in all cases where one of the scales, at most, is rectilinear. The case where two scales are rectilinear and the third is curved will be considered in Section 5; we will see that the f_i, g_i are still obtained without quadratures, simply by differentiations and eliminations. We will now turn our attention to the case where the given equation permits a nomographic representation with three rectilinear scales.

4. – The Case of $\frac{\partial^2 \log M}{\partial x \partial y} = 0$

In this case, when calculating M according to the given equation (1), we obtain an expression in the form

(65) $$M = \alpha(x)\beta(y).$$

In order to get to equation (51), by setting

(66) $$\varphi(x) = \int \frac{dx}{\alpha(x)}, \quad \psi(y) = -\int \beta(y) dy,$$

and taking φ and ψ as new independent variables, equation (1) can be reduced to an equation involving $\varphi + \psi$ and z, from which we derive

$$\varphi + \psi = -\chi(z).$$

To obtain the equation

(51) $$\varphi(x) + \psi(y) + \chi(z) = 0,$$

2. ON EQUATIONS OF THREE VARIABLES

besides the customary eliminations, we have to solve the two quadratures (66). By a change of variables, equation (51) can be reduced to the form

(67) $$x + y + z = 0.$$

We have already encountered a nomographic representation of this at the conclusion of Section 2; we shall see that others exist that are essentially different.

Equation (67) yields
$$M = -1, \quad N = 0;$$
as a result, according to (18),

(68) $$D = -C,$$

and introducing this value of D into equations (23), we obtain

(69) $$\begin{cases} 2\dfrac{\partial}{\partial y}\left(\dfrac{\partial C}{\partial x} - \dfrac{1}{2}C^2\right) - \dfrac{\partial}{\partial x}\left(\dfrac{\partial C}{\partial x} - \dfrac{1}{2}C^2\right) = 0, \\[6pt] \dfrac{\partial}{\partial y}\left(\dfrac{\partial C}{\partial y} + \dfrac{1}{2}C^2\right) - 2\dfrac{\partial}{\partial x}\left(\dfrac{\partial C}{\partial y} + \dfrac{1}{2}C^2\right) = 0. \end{cases}$$

To find all the nomographic representations of (67), we have to completely integrate this system (69). We immediately find two primary integrals, namely

(70) $$\begin{cases} \dfrac{\partial C}{\partial x} - \dfrac{1}{2}C^2 = -6\varphi(2x + y), \\[6pt] \dfrac{\partial C}{\partial y} + \dfrac{1}{2}C^2 = 6\psi(x + 2y), \end{cases}$$

such that this system is equivalent to system (69). Taking (70) into account, the condition of integratability

$$\frac{\partial^2 C}{\partial x \partial y} = \frac{\partial}{\partial y}\left(\frac{1}{2}C^2 - 6\varphi\right) = \frac{\partial}{\partial x}\left(6\psi - \frac{1}{2}C^2\right)$$

yields

(71) $$[\psi(x + 2y) - \varphi(2x + y)]\, C = \varphi'(2x + y) + \psi'(x + 2y).$$

In what follows, it is helpful to distinguish the four following cases:

    I. $\varphi' = \psi' = 0$;
    II. $\varphi' = 0 \quad \psi' \neq 0$;
    III. $\varphi' \neq 0 \quad \psi' = 0$;
    IV. $\varphi' \neq 0 \quad \psi' \neq 0$.

Case I. $\varphi' = \psi' = 0$. – We have $\varphi = \text{constant} = c_1, \psi = \text{constant} = c_2$, and for $C \neq 0$, (71) yields $c_1 = c_2$, whereas for $C = 0$, (70) yields $c_1 = c_2 = 0$.

Thus, we still have $c_1 = c_2 = -\frac{1}{3}c$; equation (71) can be reduced to an identity, and (70) becomes
$$\frac{\partial C}{\partial x} = \frac{1}{2}C^2 + 2c,$$
$$\frac{\partial C}{\partial y} = -\left(\frac{1}{2}C^2 + 2c\right),$$
from which
$$\frac{\partial C}{\partial x} + \frac{\partial C}{\partial y} = 0$$
and

(72) $$C = \chi(x-y), \quad \chi'(x-y) = \frac{1}{2}[\chi(x-y)]^2 + 2c.$$

Now, we need to subdivide our case in the following manner:
$$\mathrm{I}\alpha : c = 0; \quad \mathrm{I}\beta : c > 0; \quad \mathrm{I}\gamma : c < 0.$$

Case Iα: $c = 0$. - The equations (72) become

(73) $$C = \chi(x-y), \quad \chi' = \frac{1}{2}\chi^2.$$

Let us subdivide this case in two once again:
$$\mathrm{I}\alpha 1 : \chi = 0 \quad \text{and} \quad \mathrm{I}\alpha 2 : \chi \neq 0.$$

Case Iα1: $\chi = 0$ – Equations (73) then yield

(74) $$C = 0$$

and (68), $D = 0$; system (22) allows for the particular integral
$$\theta = 0,$$
and with this choice of θ, equations (16) and (22) have as their particular integrals
$$u = x - y, \quad v = y.$$

Calculating the f_i and g_i using formulas (11), (9), (15), and (7), we find the equation

(75) $$\begin{vmatrix} 1 & x & 1 \\ 0 & y & 1 \\ \frac{1}{2} & -\frac{1}{2}z & 1 \end{vmatrix} = \frac{1}{2}(x+y+z) = 0.$$

We should say, for the sake of brevity, that two nomograms belong to the same family if their defining equations (4) and (5) are linked by homography (6); thus, the same value of C corresponds to a family of nomograms and vice versa (fundamental theorem of Section 1), and we will designate as *parameters*

2. ON EQUATIONS OF THREE VARIABLES

of the family those arbitrary constants that enter into the expression of C. We will say, in addition, that all equations obtained by varying the parameters constitute a class.

Equation (75) shows that case Iα1 encompasses a family of nomograms with three rectilinear and concurrent scales, and without parameters. This family was discovered long ago by M. d'Ocagne.

Case Iα2: $\chi \neq 0$. – By integrating, the second equation (73) then yields

$$-\frac{1}{\chi} = \frac{1}{2}(x - y) + \text{constant}$$

or

$$-\frac{1}{\chi} = \frac{1}{2}(\overline{x - x_0} - \overline{y - y_0}),$$

where, of course, it is the difference $x_0 - y_0$ that is the arbitrary constant; the first equation (73) then shows us that

(76) $$C = \frac{2}{\overline{y - y_0} - \overline{x - x_0}}.$$

This equation indicates that scales x and y are curved; the method from Section 3 provides the corresponding nomographic equation, and we can thus limit ourselves to writing this equation and verifying that it correctly provides the value (76) for C.

By z_0, let us designate here and in the rest of this section the constant determined by the condition

(77) $$x_0 + y_0 + z_0 = 0;$$

the equation under consideration can be written as

(78) $$\begin{vmatrix} \frac{1}{x-x_0} & \frac{1}{(x-x_0)^2} & 1 \\ \frac{1}{y-y_0} & \frac{1}{(y-y_0)^2} & 1 \\ -\frac{1}{z-z_0} & 0 & 1 \end{vmatrix} = \frac{(\overline{x-x_0} - \overline{y-y_0})(x+y+z)}{(x-x_0)^2(y-y_0)^2(z-z_0)} = 0.$$

From this, with the help of (8), (16) and (17), we can successively derive that
$$u = \frac{1}{x - x_0} + \frac{1}{y - y_0},$$
$$v = -\frac{1}{(x - x_0)(y - y_0)},$$
$$e^\theta = \frac{\overline{x - x_0} - \overline{y - y_0}}{(x - x_0)^3 (y - y_0)^3},$$
$$e^\theta C = -\frac{2}{(x - x_0)^3 (y - y_0)^3},$$
whence the expression (76) for C.

According to (78), the Iα2 case gives rise to a class of nomograms where the x and y scales are supported by the same conic, the z scale being rectilinear and tangential to this conic. The families contain one parameter, namely $x_0 - y_0$. These nomograms were discovered by Clark.[6] Next we come to

Case Iβ : $c > 0$. – By letting $c = a^2$, where we may assume, for example, that $a > 0$, the equations (72) yield

(79) $$C = -2a \cot a \left(\overline{x - x_0} - \overline{y - y_0} \right),$$

and the nomographic equation becomes

(80) $$\begin{vmatrix} \cot a(x - x_0) & \cot^2 a(x - x_0) & 1 \\ \cot a(y - y_0) & \cot^2 a(y - y_0) & 1 \\ -\cot a(z - z_0) & -1 & 1 \end{vmatrix}$$
$$= \frac{\sin a \left(\overline{x - x_0} - \overline{y - y_0} \right) \sin a(x + y + z)}{\sin^2 a(x - x_0) \sin^2 a(y - y_0) \sin a(z - z_0)} = 0,$$

from which we can derive
$$u = \cot a(x - x_0) + \cot a(y - y_0),$$
$$v = -\cot(x - x_0) \cot a(y - y_0),$$
$$e^\theta = \frac{a^2 \sin a \left(\overline{x - x_0} - \overline{y - y_0} \right)}{\sin^3 a(x - x_0) \sin^3 a(y - y_0)},$$
$$e^\theta C = -\frac{2a^3 \cos a \left(\overline{x - x_0} - \overline{y - y_0} \right)}{\sin^3 a(x - x_0) \sin^3 a(y - y_0)},$$
and, as a consequence, the expression (79) for C.

---

[6] *Théorie générale des abaques d'alignement de tout ordre* (*Revue de Mécanique*, 1907). – See also: R. SOREAU, *Nouveaux types d'abaques*, etc. (*Mémoires et comptes rendus de La Société des Ingénieurs civils de France*, 1906). – D'OCAGNE, *Calcul Graphique Et Nomographie*, p. 285 and following.

2. ON EQUATIONS OF THREE VARIABLES

Equation (80) shows us that case Iβ gives rise to a class of nomograms on the x and y scales supported by the same conic, the z scale being rectilinear and not intersecting this conic. There are two parameters, $x_0 - y_0$ and a. For the special case when $a = 1$, we owe these nomograms once again to Clark.

Case Iγ: $c < 0$. – We need to subdivide this case in two:
$$I\gamma 1: \frac{1}{2}\chi^2 + 2c = 0 \quad \text{and} \quad I\gamma 2: \frac{1}{2}\chi^2 + 2c \neq 0.$$

Case Iγ1: $\frac{1}{2}\chi^2 + 2c = 0$. – By designating a as a real constant different from 0, we will then have

(81) $$C = a.$$

System (22) allows for a particular integral defined by
$$e^\theta = -a^2 e^{a(x-y)},$$
equations (16) and (21) being satisfied by setting
$$u = e^{-ay} - e^{ax}, \quad v = e^{-ay},$$
and formulas (11), (9), (15) and (7) give us the equation

(82) $$\begin{vmatrix} 0 & e^{ax} & 1 \\ 1 & e^{-ay} & 1 \\ \frac{1}{1-e^{az}} & 0 & 1 \end{vmatrix} = \frac{e^{-ay}\left(e^{a(x+y+z)} - 1\right)}{e^{az} - 1} = 0.$$

As a consequence, the Iγ1 case gives rise to a class of nomograms with three rectilinear scales that are not concurrent, a class containing a parameter a. For $a = \pm 1$, these nomograms were discovered by M. d'Ocagne, who also noted that the two families $a = 1$ and $a = -1$ are homographically different, a circumstance immediately made evident by our general theory.

Case Iγ2: $\frac{1}{2}\chi^2 + 2c \neq 0$. – Let us set $c = -a^2$, $a > 0$, and let us introduce the hyperbolic functions
$$\operatorname{sh} x = -i \sin ix,$$
$$\operatorname{ch} x = \cos ix,$$
$$\operatorname{coth} x = i \cot ix,$$
$$\dots\dots\dots\dots\dots$$

Comparison with case Iβ immediately gives us

(83) $$C = -2a \operatorname{coth} a \left(\overline{x - x_0} - \overline{y - y_0}\right),$$

(84)
$$\begin{vmatrix} \coth a(x-x_0) & \coth^2 a(x-x_0) & 1 \\ \coth a(y-y_0) & \coth^2 a(y-y_0) & 1 \\ -\coth a(z-z_0) & 1 & 1 \end{vmatrix}$$
$$= \frac{\operatorname{sh} a\,(\overline{x-x_0}-\overline{y-y_0})\,\operatorname{sh} a(x+y+z)}{\operatorname{sh}^2 a(x-x_0)\,\operatorname{sh}^2 a(y-y_0)\,\operatorname{sh} a(z-z_0)} = 0.$$

According to (84), the Iγ2 case yields a class of nomograms enclosing two parameters, $x_0 - y_0$ and a. The x and y scales are supported by the same conic, which is cut at two points by the rectilinear z scale. For $a = 1$, these nomograms were discovered by Clark.

Now we must consider

Case II: $\varphi' = 0, \chi' \neq 0$. – Since $\varphi = constant = c$, equation (71) yields
$$C = \frac{\psi'(x+2y)}{\psi(x+2y) - c};$$
now, according to (67), $x + 2y = y - z$, such that
$$C = \chi(y-z).$$

Setting y and z as independent variables, the formulas (46) show us that $C_x = C$, such that we only have to repeat the discussion for case I, using different independent variables. The Iα1 and Iγ1 cases provide us with nothing new, whereas the other cases lead to the following classes of conical nomograms:

(85)
$$\begin{vmatrix} -\frac{1}{x-x_0} & 0 & 1 \\ \frac{1}{y-y_0} & \frac{1}{(y-y_0)^2} & 1 \\ \frac{1}{z-z_0} & \frac{1}{(z-z_0)^2} & 1 \end{vmatrix} = 0, \quad C = \frac{2}{x-x_0+2(y-y_0)},$$

(86)
$$\begin{cases} \begin{vmatrix} -\cot a(x-x_0) & -1 & 1 \\ \cot a(y-y_0) & \cot^2 a(y-y_0) & 1 \\ \cot a(z-z_0) & \cot^2 a(z-z_0) & 1 \end{vmatrix} = 0, \\ C = 2a \cot a\,(\overline{x-x_0}+\overline{2y-y_0}), \end{cases}$$

and

(87)
$$\begin{cases} \begin{vmatrix} -\coth a(x-x_0) & 1 & 1 \\ \coth a(y-y_0) & \coth^2 a(y-y_0) & 1 \\ \coth a(z-z_0) & \coth^2 a(z-z_0) & 1 \end{vmatrix} = 0, \\ C = 2a \coth a\,(x-x_0+\overline{2y-y_0}). \end{cases}$$

2. ON EQUATIONS OF THREE VARIABLES

Case III: $\varphi' \neq 0, \chi' = 0$. – Analogous reasoning to what has just preceded demonstrates that by setting z and x as independent variables

$$C = C_y = \chi(z - x);$$

we obtain the new conical nomograms:

(88) $\quad \begin{vmatrix} \frac{1}{x-x_0} & \frac{1}{(x-x_0)^2} & 1 \\ -\frac{1}{y-y_0} & 0 & 1 \\ \frac{1}{z-z_0} & \frac{1}{(z-z_0)^2} & 1 \end{vmatrix} = 0, \quad C = \frac{2}{2x - x_0 + (y - y_0)},$

(89) $\quad \begin{cases} \begin{vmatrix} \cot a(x-x_0) & \cot^2 a(x-x_0) & 1 \\ -\cot a(y-y_0) & -1 & 1 \\ \cot a(z-z_0) & \cot^2 a(z-z_0) & 1 \end{vmatrix} = 0, \\ C = 2a \cot a \, (2x - x_0 + y - y_0), \end{cases}$

(90) $\quad \begin{cases} \begin{vmatrix} \coth a(x-x_0) & \coth^2 a(x-x_0) & 1 \\ -\coth a(y-y_0) & 1 & 1 \\ \coth a(z-z_0) & \coth^2 a(z-z_0) & 1 \end{vmatrix} = 0, \\ C = 2a \coth a \, (2x - x_0 + y - y_0). \end{cases}$

Finally, we come to

Case IV: $\varphi' \neq 0, \chi \neq 0$. – For the time being, letting $2x = y = \lambda, x+2y = \mu$, equation (71) yields

(91) $$C = \frac{\varphi'(\lambda) + \psi'(\mu)}{\psi(\mu) - \varphi(\lambda)},$$

and, by substituting this expression in equations (70), we obtain

(92) $\quad \begin{cases} [\psi(\mu) - \varphi(\lambda)] [2\varphi''(\lambda) + \psi''(\mu)] + \dfrac{3}{2} \left[\varphi'^2(\lambda) - \psi'^2(\mu) \right] \\ \qquad\qquad\qquad = -6\varphi(\lambda) [\psi(\mu) - \varphi(\lambda)]^2, \\ [\psi(\mu) - \varphi(\lambda)] [\varphi''(\lambda) + 2\psi''(\mu)] + \dfrac{3}{2} \left[\varphi'^2(\lambda) - \psi'^2(\mu) \right] \\ \qquad\qquad\qquad = 6\psi(\lambda) [\psi(\mu) - \varphi(\lambda)]^2. \end{cases}$

By subtracting and dividing by $\psi(\mu) - \varphi(\lambda)$, we get

$$\varphi''(\lambda) - \psi''(\mu) = 6 \left[\varphi^2(\lambda) - \psi^2(\mu) \right],$$

and as a consequence, λ and μ being independent variables,
$$\varphi''(\lambda) - 6\varphi^2(\lambda) = \text{const.} = -\frac{1}{2}g_2,$$
$$\psi''(\mu) - 6\psi^2(\mu) = -\frac{1}{2}g_2.$$

Multiplying the first of these equations by $2\varphi'(\lambda)$, the second by $2\psi'(\mu)$ and integrating, we obtain

(93)
$$\begin{cases} \varphi'^2(\lambda) = 4\varphi^3(\lambda) - g_2\varphi(\lambda) - g_3, \\ \psi'^2(\mu) = 4\psi^3(\mu) - g_2\psi(\mu) - g'_3, \end{cases}$$

where g_3 and g'_3 are new constants, and by substituting the expressions (93) in any of equations (92), the former may be reduced to

(94) $$g'_3 = g_3.$$

Introducing Weierstrass's elliptical function $\wp\mu$, we see that in consideration of (94), the general integrals of equations (93) become
$$\varphi(\lambda) = \wp(\lambda - \lambda_0; g_2, g_3),$$
$$\psi(\mu) = \wp(\mu - \mu_0; g_2, g_3),$$
where λ_0 and μ_0 are arbitrary constants. By setting $\lambda_0 = 2x_0 + y_0$, and $\mu_0 = x_0 + 2y_0$, we obtain from (91)

(95) $$C = \frac{\wp'(x - x_0 + 2\overline{y - y_0}; g_2, g_3) + \wp'(2\overline{x - x_0} + y - y_0; g_2, g_3)}{\wp(x - x_0 + 2\overline{y - y_0}; g_2, g_3) - \wp(2\overline{x - x_0} + y - y_0; g_2, g_3)},$$

an expression containing four arbitrary constants, x_0, y_0, g_2, g_3.

By virtue of (67), the preceding expression may also be written as
$$C = \frac{\wp'(y - y_0 - \overline{z - z_0}) - \wp'(z - z_0 - \overline{x - x_0})}{\wp(y - y_0 - \overline{z - z_0}) - \wp(z - z_0 - \overline{x - x_0})},$$
and by applying the well-known formula

(96) $$\frac{1}{2}\frac{\wp'u - \wp'v}{\wp u - \wp v} = \frac{\tau'}{\tau}(u + v) - \frac{\tau'}{\tau}u - \frac{\tau'}{\tau}v,$$

this yields

(97) $$C = -2\left[\frac{\tau'}{\tau}(x - x_0 - \overline{y - y_0}) + \frac{\tau'}{\tau}(y - y_0 - \overline{z - z_0}) + \frac{\tau'}{\tau}(z - z_0 - \overline{x - x_0})\right]$$

2. ON EQUATIONS OF THREE VARIABLES

or, by applying (96) once again,

(98)
$$C = -\frac{\wp'(x-x_0)+\wp'(y-y_0)}{\wp(x-x_0)-\wp(y-y_0)} - \frac{\wp'(y-y_0)+\wp'(z-z_0)}{\wp(y-y_0)-\wp(z-z_0)} - \frac{\wp'(z-z_0)+\wp'(x-x_0)}{\wp(z-z_0)-\wp(x-x_0)},$$

formulas that are symmetrical for x, y, z.

The nomographic equation corresponding to this value of C is

(99)
$$\begin{vmatrix} \wp(x-x_0) & \wp'(x-x_0) & 1 \\ \wp(y-y_0) & \wp'(y-y_0) & 1 \\ \wp(z-z_0) & \wp'(z-z_0) & 1 \end{vmatrix} = 0.$$

First of all, this equation is actually equivalent to (67), by virtue of the well-known formula

$$\begin{vmatrix} \wp u & \wp' u & 1 \\ \wp v & \wp' v & 1 \\ \wp w & \wp' w & 1 \end{vmatrix} = \frac{2\tau(u-v)\tau(v-w)\tau(w-u)\tau(u+v+w)}{\tau^3 u \tau^3 v \tau^3 w},$$

and secondly, we will verify in the following manner that it leads to the value (98) for C. The equations (8) and (9) yield

(100)
$$\begin{aligned} u &= \frac{\wp'(x-x_0)-\wp'(y-y_0)}{\wp(x-x_0)-\wp(y-y_0)} \\ &= 2\frac{\tau'}{\tau}(x-x_0+y-y_0) - 2\frac{\tau'}{\tau}(x-x_0) - 2\frac{\tau'}{\tau}(y-y_0) \\ &= -2\left[\frac{\tau'}{\tau}(x-x_0) + \frac{\tau'}{\tau}(y-y_0) + \frac{\tau'}{\tau}(z-z_0)\right], \\ v &= \wp'(x-x_0) - u\wp(x-x_0) = \wp'(y-y_0) - u\wp(y-y_0), \end{aligned}$$

from which

(101)
$$\begin{cases} \dfrac{\partial u}{\partial x} = 2[\wp(x-x_0)-\wp(z-z_0)], & \dfrac{\partial u}{\partial y} = 2[\wp(y-y_0)-\wp(z-z_0)], \\ \dfrac{\partial v}{\partial x} = -\dfrac{\partial u}{\partial x}\wp(y-y_0), & \dfrac{\partial v}{\partial y} = -\dfrac{\partial u}{\partial y}\wp(x-x_0), \end{cases}$$

(102)
$$\begin{aligned} e^\theta = 4[\wp(x-x_0)-\wp(y-y_0)][\wp(y-y_0)-\wp(z-z_0)] \\ \times [\wp(z-z_0)-\wp(x-x_0)], \end{aligned}$$

(103) $$\begin{cases} \dfrac{\partial^2 u}{\partial x^2} = 2[\wp'(x-x_0) - \wp'(z-z_0)] & = -\dfrac{\wp'(z-z_0) + \wp'(x-x_0)}{\wp(z-z_0) - \wp(x-x_0)} \dfrac{\partial u}{\partial x}, \\ \dfrac{\partial^2 v}{\partial x^2} = -\dfrac{\partial^2 u}{\partial x^2} \wp(y-y_0) & = -\dfrac{\wp'(z-z_0) + \wp'(x-x_0)}{\wp(z-z_0) - \wp(x-x_0)} \dfrac{\partial v}{\partial x} \end{cases}$$

and finally,

(104) $$\begin{cases} \dfrac{\partial^2 u}{\partial x \partial y} = 2\wp'(z-z_0), \\ \dfrac{\partial^2 v}{\partial x \partial y} = -\dfrac{\partial^2 u}{\partial x \partial y} \wp(x-x_0) - \dfrac{\partial u}{\partial y} \wp'(x-x_0). \end{cases}$$

From (103) and (101), we obtain

(105) $$\begin{aligned} & \left(\dfrac{\partial^2 u}{\partial x^2} \dfrac{\partial v}{\partial y} - \dfrac{\partial^2 v}{\partial x^2} \dfrac{\partial u}{\partial y} \right) e^{-\theta} \\ & = -\dfrac{\wp'(z-z_0) + \wp'(x-x_0)}{\wp(z-z_0) - \wp(x-x_0)} \left(\dfrac{\partial u}{\partial x} \dfrac{\partial v}{\partial y} - \dfrac{\partial u}{\partial y} \dfrac{\partial v}{\partial x} \right) e^{-\theta} \\ & = -\dfrac{\wp'(z-z_0) + \wp'(x-x_0)}{\wp(z-z_0) - \wp(x-x_0)}, \end{aligned}$$

and from (104) and (101) we obtain

(106) $$\begin{aligned} & 2\left(\dfrac{\partial^2 u}{\partial x \partial y} \dfrac{\partial v}{\partial x} - \dfrac{\partial^2 v}{\partial x \partial y} \dfrac{\partial u}{\partial x} \right) e^{-\theta} \\ & = 2\left\{ \dfrac{\partial^2 u}{\partial x \partial y} \dfrac{\partial u}{\partial x} [\wp(x-x_0) - \wp(y-y_0)] + \dfrac{\partial u}{\partial x} \dfrac{\partial u}{\partial y} \wp'(x-x_0) \right\} e^{-\theta} \\ & = -\dfrac{2\wp'(z-z_0)}{\wp(y-y_0) - \wp(z-z_0)} - \dfrac{2\wp'(x-x_0)}{\wp(x-x_0) - \wp(y-y_0)} \\ & = -\dfrac{\wp'(y-y_0) + \wp'(z-z_0)}{\wp(y-y_0) - \wp(z-z_0)} + \dfrac{\wp'(y-y_0) - \wp'(z-z_0)}{\wp(y-y_0) - \wp(z-z_0)} \\ & \quad - \dfrac{\wp'(x-x_0) + \wp'(y-y_0)}{\wp(x-x_0) - \wp(y-y_0)} - \dfrac{\wp'(x-x_0) - \wp'(y-y_0)}{\wp(x-x_0) - \wp(y-y_0)} \\ & = -\dfrac{\wp'(y-y_0) + \wp'(z-z_0)}{\wp(y-y_0) - \wp(z-z_0)} - \dfrac{\wp'(x-x_0) - \wp'(y-y_0)}{\wp(x-x_0) - \wp(y-y_0)} \end{aligned}$$

since equation (99) may be written as

$$\dfrac{\wp'(y-y_0) - \wp'(z-z_0)}{\wp(y-y_0) - \wp(z-z_0)} = \dfrac{\wp'(x-x_0) - \wp'(y-y_0)}{\wp(x-x_0) - \wp(y-y_0)}.$$

By taking the sum of expressions (105) and (106), we thus obtain, by virtue of (17), the value (98) of C.

2. ON EQUATIONS OF THREE VARIABLES

We can see that case IV gives rise to a class of nomograms with four parameters x_0, y_0, g_2, g_3, where the three scales are supported by the same genus 1 cubic, transformed homographically into the curve

$$\eta^2 = 4\xi^3 - g_2\xi - g_3.$$

The specific cases where the elliptical function $\wp u$ degenerates deserve particular mention. They are characterized by the equation

$$g_2^3 - 27g_3^2 = 0,$$

which leads to a division into three cases.

First, assuming that $g_2 > 0, g_3 > 0$, we will have

$$g_2 = \frac{4}{3}a^4, \quad g_3 = \frac{8}{27}a^6 \quad (a > 0),$$

$$\wp u = \frac{a^2}{\sin^2 au} - \frac{a^2}{3},$$

and equation (99) becomes

(107) $\quad\begin{vmatrix} \frac{1}{\sin^2 a(x-x_0)} & \frac{\cos a(x-x_0)}{\sin^3 a(x-x_0)} & 1 \\ \frac{1}{\sin^2 a(y-y_0)} & \frac{\cos a(y-y_0)}{\sin^3 a(y-y_0)} & 1 \\ \frac{1}{\sin^2 a(z-z_0)} & \frac{\cos a(z-z_0)}{\sin^3 a(z-z_0)} & 1 \end{vmatrix} = 0.$

Equation (107) defines a class of nomograms with three parameters, a, x_0, y_0, whose three scales are supported by the same cubic of genus 0, a homographic of

$$\eta^2 = \xi^2(\xi - 1),$$

and having, as a consequence, an isolated point.

Assuming, secondly, that $g_2 > 0$, $g_3 < 0$, we will have

$$g_2 = \frac{4}{3}a^4, \quad g_3 = -\frac{8}{27}a^6 \quad (a > 0),$$

$$\wp u = \frac{a^2}{\text{sh}^2 au} + \frac{a^2}{3},$$

and equation (99) becomes

(108) $\quad\begin{vmatrix} \frac{1}{\text{sh}^2 a(x-x_0)} & \frac{\text{ch} a(x-x_0)}{\text{sh}^3 a(x-x_0)} & 1 \\ \frac{1}{\text{sh}^2 a(y-y_0)} & \frac{\text{ch} a(y-y_0)}{\text{sh}^3 a(y-y_0)} & 1 \\ \frac{1}{\text{sh}^2 a(z-z_0)} & \frac{\text{ch} a(z-z_0)}{\text{sh}^3 a(z-z_0)} & 1 \end{vmatrix} = 0.$

Equation (108) defines a class of nomograms with three parameters, a, x_0, y_0, whose three scales are supported by the same cubic of type 0, a homographic of
$$\eta^2 = \xi^2(\xi+1),$$
and having, as a consequence, a double point with distinct tangents. Finally, assuming that $g_2 = g_3 = 0$, we will have
$$\wp u = \frac{1}{u^2},$$
and equation (99) becomes

(109)
$$\begin{vmatrix} \frac{1}{(x-x_0)^2} & \frac{1}{(x-x_0)^3} & 1 \\ \frac{1}{(y-y_0)^2} & \frac{1}{(y-y_0)^3} & 1 \\ \frac{1}{(z-z_0)^2} & \frac{1}{(z-z_0)^3} & 1 \end{vmatrix} = 0.$$

This equation defines a class of nomograms with two parameters, x_0, y_0, whose three scales are supported by the same cubic of type 0, a homographic of
$$\eta^2 = \xi^3,$$
and having, as a consequence, a double point with common tangents.

The nomograms (107), (108), for $a = 1$, and (109) were discovered by Clark (*loc. cit.*).

5. – The Case of Two Rectilinear Scales, and the Third One Being Curved

By appropriately choosing independent variables, we can assume that the rectilinear scales are those of x and y.

As we have seen in Section 2, the necessary and sufficient condition for them to be rectilinear is expressed by the equations
$$D = MC + N,$$
$$\frac{\partial C}{\partial y} = \frac{\partial D}{\partial x} = 0.$$

By substituting the value D in the last equation, it becomes
$$M\frac{\partial C}{\partial x} + \frac{\partial M}{\partial x}C + \frac{\partial N}{\partial x} = 0$$
or
$$\frac{\partial C}{\partial x} = -\frac{\partial \log M}{\partial x}C - \frac{1}{M}\frac{\partial N}{\partial x}.$$

By virtue of $\frac{\partial C}{\partial y} = 0$, we now have

$$0 = \frac{\partial^2 C}{\partial x \partial y} = -\frac{\partial}{\partial y}\left(\frac{\partial \log M}{\partial x}C + \frac{1}{M}\frac{\partial M}{\partial x}\right) = -\frac{\partial^2 \log M}{\partial x \partial y}C - \frac{\partial}{\partial y}\frac{1}{M}\frac{\partial M}{\partial x},$$

from which

(110) $$C = -\frac{\frac{\partial}{\partial y}\frac{1}{M}\frac{\partial N}{\partial x}}{\frac{\partial^2 \log M}{\partial x \partial y}},$$

an expression whose denominator does not cancel out identically, since if it did, the z scale would also be rectilinear by virtue of Section 4.

In order that the x and y scales be rectilinear, and the z scale be curved, it is thus necessary and sufficient that expression (110) satisfy the equations

(111) $$\frac{\partial C}{\partial y} = \frac{\partial}{\partial x}(MC + N) = 0.$$

This condition was derived in an entirely different fashion by Massau.

To return from expression (110) to the f_i, g_i, Massau provides formulas that require four quadratures and, since that time, Lecornu provided other formulas that only involve three.[7]

I will now demonstrate how the f_i, g_i can be obtained without the need for any quadratures.

Using a suitable homographic transformation, we can make the x and y scales coincide with the ξ and η axes, respectively, such that $g_1(x) = f_2(y) = 0$. Equation (4) then becomes

(112) $$\begin{vmatrix} f_1(x) & 0 & 1 \\ 0 & g_2(y) & 1 \\ f_3(z) & g_3(z) & 1 \end{vmatrix} = 0$$

or

(113) $$f_3 g_2 + f_1 g_3 - f_1 g_2 = 0.$$

---

[7]See D'OCAGNE, Traité de Nomographie, p. 422–427.

From this, we successively derive

(114)
$$\begin{cases} \dfrac{\partial z}{\partial x} = -\dfrac{(g_3 - g_2)f_1'}{f_3'g_2 + g_3'f_1}, \\ \dfrac{\partial z}{\partial y} = -\dfrac{(f_3 - f_1)g_2'}{f_3'g_2 + g_3'f_1}, \\ M = -\dfrac{f_3 - f_1}{g_3 - g_2}\dfrac{g_2'}{f_1'} \end{cases}$$

and

(115)
$$\begin{cases} \dfrac{\partial \log M}{\partial x} = -\dfrac{f_3'g_3 - g_3'(f_3 - 2f_1)}{(f_3 - f_1)(f_3'g_2 + g_3'f_1)} - \dfrac{f_1''}{f_1'}, \\ \dfrac{\partial \log M}{\partial y} = \dfrac{g_3'f_3 - f_3'(g_3 - 2g_2)}{(g_3 - g_2)(f_3'g_2 + g_3'f_1)} - \dfrac{g_2''}{g_2'}. \end{cases}$$

Subsequently, we obtain

$$\frac{\partial^2 \log M}{\partial x \partial y} = \frac{f_1'g_2'[g_3f_1 + g_2(f_3 - 2f_1)](f_3''g_3' - g_3''f_3')}{(f_3'g_2 + g_3'f_1)^3},$$

or, by virtue of (113),

(116)
$$\frac{\partial^2 \log M}{\partial x \partial y} = -\frac{f_1g_2 f_1'g_2'(f_3''g_3' - g_3''f_3')}{(f_3'g_2 + g_3'f_1)^3}.$$

Equations (114) and (116) yield

$$\frac{\left(\frac{\partial z}{\partial x}\right)^2 \frac{\partial z}{\partial y}}{\frac{\partial^2 \log M}{\partial x \partial y}} = \frac{(g_3 - g_2)^2(f_3 - f_1)}{f_1g_2(f_3''g_3' - g_3''f_3')}f_1';$$

now, according to (113),

$$g_3 - g_2 = -\frac{f_3 g_2}{f_1}, \qquad f_3 - f_1 = -\frac{f_1 g_3}{g_2},$$

such that

(117)
$$\frac{\left(\frac{\partial z}{\partial x}\right)^2 \frac{\partial z}{\partial y}}{\frac{\partial^2 \log M}{\partial x \partial y}} = -\frac{f_1'}{f_1^2}\frac{f_3^2 g_3}{f_3''g_3' - g_3''f_3'},$$

and, in the same way, we obtain

(118)
$$\frac{\frac{\partial z}{\partial x}\left(\frac{\partial z}{\partial y}\right)^2}{\frac{\partial^2 \log M}{\partial x \partial y}} = -\frac{g_2'}{g_2^2}\frac{f_3 g_3^2}{f_3''g_3' - g_3''f_3'}.$$

For the sake of brevity, we will designate by the notations $(\Phi(x, y, z))_x$ and $(\Phi(x, y, z))_y$ the results we obtain by eliminating the variables x and y, respectively, from $\Phi(x, y, z)$, with the help of the given equation (1).

2. ON EQUATIONS OF THREE VARIABLES

Having set this, equations (117) and (118) inform us that

(119)
$$\begin{cases} \left[\dfrac{\left(\frac{\partial z}{\partial x}\right)^2 \frac{\partial z}{\partial y}}{\frac{\partial^2 \log M}{\partial x \partial y}} \right]_y = \varphi(x)\chi_1(z), \\ \left[\dfrac{\frac{\partial z}{\partial x}\left(\frac{\partial z}{\partial y}\right)^2}{\frac{\partial^2 \log M}{\partial x \partial y}} \right]_x = \psi(y)\chi_2(z), \end{cases}$$

where, of course, φ, ψ, χ_1, and χ_2 are obtained from the given equation by means of differentiation and elimination. Comparison of (117), (118), and (119), yields

(120)
$$\begin{cases} f_1'' = \varphi f_1^2, & g_2' = \psi g_2^2, \\ \dfrac{f_3^2 g_3}{f_3'' g_3' - g_3'' f_3'} = -\chi_1, & \dfrac{f_3 g_3^2}{f_3'' g_3' - g_3'' f_3'} = -\chi_2. \end{cases}$$

Let us note in passing that the most general decomposition into factors of the left elements of (119) can be obtained by replacing φ, ψ, χ_1, and χ_2 with

$$\bar\varphi = \frac{1}{c_1}\varphi, \quad \bar\chi_1 = c_1\chi_1,$$

$$\bar\psi = \frac{1}{c_2}\psi, \quad \bar\chi_2 = c_2\chi_2,$$

c_1 and c_2 being any two constants. Now, according to (120), this comes back to replacing f_1, g_2, f_3, g_3, by $c_1 f_1$, $c_2 g_2$, $c_1 f_3$, $c_2 g_3$, respectively, a homography that leaves (112) unchanged.

From equation (113), we can draw

$$g_2 = -\frac{f_1 g_3}{f_3 - f_1},$$

and by substituting this in the first of equations (114),

$$\frac{\partial z}{\partial x} = \frac{f_3 g_3}{f_3' g_3 - g_3'(f_3 - f_1)} \frac{f_1'}{f_1}.$$

By using the formulas

$$g_3 - g_2 = -\frac{f_3 g_2}{f_1}, \quad f_3 - f_1 = -\frac{f_1 g_3}{g_2},$$

the last equation (114) yields

$$M = -\frac{f_1^2 g_3}{g_2^2 f_3} \frac{g_2'}{f_1'},$$

from which, by using the value of $\frac{\partial z}{\partial x}$ that we have just given,

$$\frac{\partial \log M}{\partial x} = \left(\frac{g_3'}{g_3} - \frac{f_3'}{f_3}\right) \frac{f_3 g_3}{f_3' g_3 - g_3'(f_3 - f_1)} \frac{f_1'}{f_1} + \frac{2 f_1'}{f_1} - \frac{f_1''}{f_1'}$$

or

$$\frac{\partial \log M}{\partial x} = -\frac{f_3' g_3 - g_3' f_3}{f_3' g_3 - g_3'(f_3 - f_1)} \frac{f_1'}{f_1} + \frac{2 f_1'}{f_1} - \frac{f_1''}{f_1'}.$$

Now, the first equation (120) yields

$$\frac{f_1'}{f_1} = \varphi f_1, \quad \frac{f_1''}{f_1'} - \frac{2 f_1'}{f_1} = \frac{\varphi'}{\varphi},$$

such that the preceding equation becomes

$$\left(\frac{\partial \log M}{\partial x}\right)_y = -\frac{(f_3' g_3 - g_3' f_3)\varphi f_1}{f_3' g_3 - g_3' f_3 + g_3' f_1} - \frac{\varphi'}{\varphi},$$

or

(121) $$-\frac{\varphi}{\frac{\varphi'}{\varphi} + \left(\frac{\partial \log M}{\partial x}\right)_y} = \frac{1}{f_1} + \frac{g_3'}{f_3' g_3 - g_3' f_3},$$

and in an entirely analogous manner, one finds that

(122) $$\frac{\psi}{\left(\frac{\partial \log M}{\partial y}\right)_x - \frac{\psi'}{\psi}} = \frac{1}{g_2} + \frac{g_3'}{f_3' g_3 - g_3' f_3}.$$

In equation (121), the right element is the sum of a function of x and a function of z; (121) subsequently determines $\frac{1}{f_1}$ as a nearby constant additive, and the same remark applies to the determination of $\frac{1}{g_2}$ by (122). If we let $\bar{f}_1$ and $\bar{g}_2$ be two solutions to the equations (121) (122); we will have

$$\frac{1}{\bar{f}_1} = \frac{1}{f_1} + c_1, \quad \frac{1}{\bar{g}_2} = \frac{1}{g_2} + c_2,$$

and as a consequence, we will pass from one solution to the other by applying the homographic transformation

$$\bar{f}_i = \frac{f_i}{c_1 f_i + c_2 y_i + 1}, \quad \bar{g}_i = \frac{g_i}{c_1 f_i + c_2 g_i + 1} \quad (i = 1, 2, 3)$$

to equation (112).

In conclusion, we have developed the following method for calculating the f_i, g_i: let us first effectuate decomposition (119) by whatever means, and then let us take for f_1 and g_2 any solutions of (121) and (122) whatsoever; then, let us set $f_2 = g_1 = 0$ and finally, let us calculate f_3 and g_3 using formulas (15) and (7).

6. – Clark's Conical Nomograms

In these nomograms, two of the scales are supported by the same conic, and the third one is of any kind; by choosing the independent variables appropriately, we can act in such a way as to make the x and y scales situated upon the same conic. Using a suitable homography, we can transform this conic into the parabola

$$\eta = \xi^2,$$

and equation (4) now will take the form

(123)
$$\begin{vmatrix} f_1(x) & f_1^2(x) & 1 \\ f_2(y) & f_2^2(y) & 1 \\ f_3(z) & g_3(z) & 1 \end{vmatrix} = 0.$$

First we will search for a necessary condition for a given equation (1) to be reducible to the form (123) and subsequently show that this condition is also sufficient.

Beginning from (123), the formulas in Section 1 successively give us

(124)
$$\begin{cases} u = f_1 + f_2, \\ v = -f_1 f_2, \\ e^\theta = f_1' f_2'(f_2 - f_1), \\ C = \dfrac{f_1''}{f_1'} + \dfrac{2 f_1'}{f_2 - f_1}, \\ D = \dfrac{f_2''}{f_2'} - \dfrac{2 f_2'}{f_2 - f_1}, \end{cases}$$

from which we can immediately draw the equation

(125)
$$\frac{\partial C}{\partial y} = \frac{\partial D}{\partial x} = -\frac{2 f_1' f_2'}{(f_2 - f_1)^2} \neq 0.$$

Substituting $\frac{\partial D}{\partial x} = \frac{\partial C}{\partial y}$ into equations (23), these become

(126)
$$\begin{cases} \dfrac{\partial^2 C}{\partial x \partial y} = C \dfrac{\partial C}{\partial y}, \\ \dfrac{\partial^2 C}{\partial y^2} = D \dfrac{\partial C}{\partial y}. \end{cases}$$

Finally, eliminating D with the help of (18), we will have

(127)
$$\begin{cases} \dfrac{\partial C}{\partial y} = \dfrac{\partial}{\partial x}(MC+N), \\ \dfrac{\partial^2 C}{\partial x \partial y} = C\dfrac{\partial C}{\partial y}, \\ \dfrac{\partial^2 C}{\partial y^2} = (MC+N)\dfrac{\partial C}{\partial y}. \end{cases}$$

The first of these equations yields

(128)
$$\dfrac{\partial C}{\partial y} = M\dfrac{\partial C}{\partial x} + \dfrac{\partial M}{\partial x}C + \dfrac{\partial N}{\partial x},$$

the second

(129)
$$\dfrac{\partial^2 C}{\partial x \partial y} = C\left(M\dfrac{\partial C}{\partial x} + \dfrac{\partial M}{\partial x}C + \dfrac{\partial N}{\partial x}\right),$$

and the third, by substituting in it the value (128) of $\dfrac{\partial C}{\partial y}$,

$$M\dfrac{\partial^2 C}{\partial x \partial y} + \dfrac{\partial M}{\partial y}\dfrac{\partial C}{\partial x} + \dfrac{\partial M}{\partial x}\dfrac{\partial C}{\partial y} + \dfrac{\partial^2 M}{\partial x \partial y}C + \dfrac{\partial^2 N}{\partial x \partial y}$$
$$= (MC+N)\left(M\dfrac{\partial C}{\partial x} + \dfrac{\partial M}{\partial x}C + \dfrac{\partial N}{\partial x}\right),$$

or by replacing $\dfrac{\partial^2 C}{\partial x \partial y}$ and $\dfrac{\partial C}{\partial y}$ by their values (129) and (128) and reducing, while taking into account the relation (14) between M and N,

$$\left(\dfrac{\partial^2 M}{\partial x \partial y} - \dfrac{1}{M}\dfrac{\partial M}{\partial x}\dfrac{\partial M}{\partial y}\right)C + \dfrac{\partial^2 N}{\partial x \partial y} - \dfrac{1}{M}\dfrac{\partial M}{\partial y}\dfrac{\partial N}{\partial x} = 0,$$

and this equation yields

(130)
$$C = -\dfrac{\dfrac{\partial}{\partial y}\dfrac{1}{M}\dfrac{\partial N}{\partial x}}{\dfrac{\partial^2 \log M}{\partial x \partial y}}.$$

In this expression, the denominator does not cancel identically, for otherwise we would have the case described in Section 4, and the z scale would be rectilinear.

According to the calculations that we have just completed, *it is necessary, in order that the given equation* (1) *be reducible to the form* (123), *that the expression* (130) *for C satisfy the equations*

(131)
$$\begin{cases} \dfrac{\partial C}{\partial y} = \dfrac{\partial}{\partial x}(MC+N) = \dfrac{\partial D}{\partial x} \neq 0, \\ \dfrac{\partial^2 C}{\partial x \partial y} = C\dfrac{\partial C}{\partial y}. \end{cases}$$

2. ON EQUATIONS OF THREE VARIABLES

I would state that *this condition is also sufficient.*

To demonstrate this, we must find two functions $f_1(x)$ and $f_2(y)$ satisfying the two last equations in (124); the method from Section 3 can serve to accomplish this here. But the final formula being of unexpected simplicity, it would make more sense to demonstrate this directly.

Let us designate by y_0 a constant such that $\frac{\partial C}{\partial y} \neq 0$ for $y = y_0$,[8] and let us write, for the sake of brevity

$$C_0 = C(x, y_0),$$

$$\left(\frac{\partial C}{\partial x}\right)_0 = \left[\frac{\partial C(x,y)}{\partial x}\right]_{y=y_0},$$

$$\left(\frac{\partial C}{\partial y}\right)_0 = \left[\frac{\partial C(x,y)}{\partial y}\right]_{y=y_0}, \text{ etc.}$$

By integrating with respect to y, the last equation (131) yields

(132)
$$\frac{\partial C}{\partial x} - \left(\frac{\partial C}{\partial x}\right)_0 = \frac{1}{2}(C^2 - C_0^2).$$

Let us now consider the function

(133)
$$\varphi(x,y) = \frac{\left(\frac{\partial C}{\partial y}\right)_0}{C_0 - C};$$

we have

$$\frac{\partial \varphi(x,y)}{\partial x} = \frac{C_0 \left(\frac{\partial^2 C}{\partial x \partial y}\right)_0}{C_0 - C} - \frac{\left(\frac{\partial C}{\partial y}\right)_0 \left[\left(\frac{\partial C}{\partial x}\right)_0 - \frac{\partial C}{\partial x}\right]}{(C_0 - C)^2},$$

or, by virtue of (131) and (132),

$$\frac{\partial \varphi(x,y)}{\partial x} = \frac{C_0 \left(\frac{\partial C}{\partial y}\right)_0}{C_0 - C} - \frac{1}{2} \frac{\left(\frac{\partial C}{\partial y}\right)_0 (C_0^2 - C^2)}{(C_0 - C)^2}$$

or, finally

(134)
$$\frac{\partial \varphi(x,y)}{\partial x} = \frac{1}{2}\left(\frac{\partial C}{\partial y}\right)_0.$$

---

[8] This choice of the constant y_0 is possible; because if $\frac{\partial C}{\partial y} = \frac{\partial D}{\partial x} = 0$ we would be in the situation of section 5 where the x and the y axes are rectilinear since expression (130) for C coincides with the expression (110), which is valid for the example indicated.

This equation clearly yields
$$\frac{\partial^2 \varphi(x,y)}{\partial x \partial y} = 0.$$

As a result, by setting

(135) $$f_1(x) - f_2(y) = \frac{\left(\frac{\partial C}{\partial y}\right)_0}{C_0 - C},$$

this equation determines f_1 and f_2 to the same nearby constant additive, and I claim that *the functions f_1 and f_2 thus determined can figure in an equation (128) equivalent to the given equation (1)*. To demonstrate this, we first have to see that they satisfy (124). The equations (135), (134) and (131) yield

$$f_1'(x) = \frac{1}{2}\left(\frac{\partial C}{\partial y}\right)_0, \quad f_1'' = \frac{1}{2}\left(\frac{\partial^2 C}{\partial x \partial y}\right)_0 = \frac{1}{2}C_0\left(\frac{\partial C}{\partial y}\right)_0,$$

from which

$$\frac{f_1''}{f_1'} + \frac{2f_1'}{f_2 - f_1} = C_0 + (C - C_0) = C;$$

furthermore, (135) gives

$$f_2'(y) = -\frac{\left(\frac{\partial C}{\partial y}\right)_0 \frac{\partial C}{\partial y}}{(C_0 - C)^2},$$

and since $\frac{\partial^2 C}{\partial y^2} = D\frac{\partial C}{\partial y}$ according to the last equation (126) which is a consequence of (130) and (131),

$$\frac{d}{dy}\log f_2'(y) = D + \frac{2\frac{\partial C}{\partial y}}{C_0 - C},$$

such that

$$\frac{f_2''}{f_2'} - \frac{2f_2'}{f_2 - f_1} = \left(D + \frac{2\frac{\partial C}{\partial y}}{C_0 - C}\right) - \frac{2\frac{\partial C}{\partial y}}{C_0 - C} = D.$$

Equations (124) are thus satisfied, and, as a result, equations (8), (10), (12), (18), (21), (22), and (23) as well. The formulas (9) yield

$$g_1(x) = f_1^2(x),$$
$$g_2(y) = f_2^2(y),$$

and reduction of the equation given in the form (123) is completed by determining $f_3(z)$ and $g_3(z)$ by formulas (15) and (7).

CHAPTER 3

Commentary

RON DOERFLER

Among important works in mathematics, there are those that provide a real sense of the vision of the artist, a strikingly clear understanding of a field of study. T. H. Gronwall's article is not one of them. The ever-so-brief introduction; the relentless, methodical equations marching along in careful and determined order; the lack of a single example or figure; and the abrupt ending—all of these give the sense that we are missing the man behind the machine. Surely there is something beautiful behind the graceful curves of the partial derivatives here. What was in Gronwall's mind? What was he seeing, and what was he trying to accomplish? The first chapter of this volume places Gronwall's contributions within the historical arc of the field of nomography. Here we take a detailed look at the mathematics within the article to understand the ideas behind it—to see what Gronwall was seeing.

We are fortunate in our subject. The field of nomography is one of the most interactive and visually engaging in all of mathematics. Another form of graphical calculator is the sundial, and like a sundial a nomogram can be appreciated by anyone with the briefest knowledge of what it's all about. Like a sundial, a nomogram is an interactive device that is simple to use once the hard part—the dedicated work to create the design—is finished. And like a sundial, the deftness of the artisan is often accompanied by the inspiration of the artist.

In his paper, Gronwall proceeds to lay out necessary and sufficient conditions for an equation $f(x, y, z) = 0$ to be represented by an alignment nomogram. His larger purpose is to identify the types of nomograms that can be created by such an equation. This leads to a natural categorization of the *families* of three-variable nomograms and the *classes* of equations that are associated with *homographic* transformations of the nomograms. Gronwall describes nomograms of the various kinds without illustrating them, an omission we will attempt to rectify here. There is some irony in the fact that the same computers that rendered many nomograms obsolete are excellent tools for creating them [**276**]. You are encouraged to use the edge of a sheet of paper to verify that these nomograms really do work. Nomograms manifest a surprising degree of complexity in their apparent simplicity. The more they are used, the more impressive they become.

This presentation generally follows the order of presentation in Gronwall's paper, although connections to other sections are described as they occur. To make it easier to align the two presentations, equations in this chapter that also appear in Gronwall's paper are numbered in accordance with their corresponding numbers in Gronwall's paper. Gronwall did likewise for equations repeated in his paper. Unique equations in this chapter are numbered as they occur. Note also that the nomograms shown here are naturally small in size; for practical use nomograms are printed in greater detail on larger sheets of paper for better precision, generally to three significant digits.

Before we begin, we should understand the basic mathematical subject of nomograms. The next section offers a brief introduction to nomography that provides a background sufficient for our study of Gronwall's paper.

3.1. The Design of Nomograms

As described in the first chapter of this volume, a simple nomogram consists of a set of scales, one per variable, laid out in such a way that a line crossing them will connect variable values that are related by an equation. This line is called an *isopleth* or *index line*, and a straightedge is generally used rather than actually drawing a line. A compound nomogram is used when it is not possible to line up all the scales for a single-line solution; here the nomogram is broken into two or more simple nomograms sharing a common scale, so that the termination of an isopleth at the shared scale becomes the start of another isopleth through the next stage. In this way nomograms provide very fast results for equations that can be very complicated. A slide rule requires multiple operations to solve many kinds of equations; a nomogram is designed to solve a specific equation very quickly.

We can solve an equation of n variables for an unknown variable by connecting the scale values of the other variables and reading the value from the scale for the unknown variable. Beyond that, the graphical layout of the scales provides a visual model of the behavior of an equation. Relationships between variables and limits on combinations of values in the solution of the equation are apparent, as well as the sensitivities of changes to variables in different regions. There is no comparable way of visualizing a multi-dimensional equation; this is a unique property of nomograms even in an age when analysts use computers to create individual 2D or 3D slices of higher-dimensional relationships. Nomograms can also be used to solve for a variable that cannot be algebraically isolated in an equation, something that requires an iterative program on a computer.

Nomograms were historically used in many areas of applied mathematics and engineering. We have already been introduced to their use in ballistics by Gronwall and others, an application that required fast computations before fire control systems were available. A sampling of other areas includes civil, electrical, mechanical and aeronautical engineering, as well as astronomy, statistics and operations research. Nomograms are still used for applications that benefit from fast calculations where three-digit accuracy is sufficient.

Beyond that, nomography comprises a very interesting field of mathematical study. It is not at all obvious how one finds a layout of scales that solves an equation, much less optimizes the layout for the greatest practical precision. In fact, not all equations can be represented by a nomogram—many cannot. This existence property is a problem that provoked a great deal of work by mathematicians [**269**], including Gronwall in the paper presented here [**269**, pp. 173–178].

Nonetheless, practical approaches for designing nomograms were developed, and standard forms of nomograms were derived for a variety of different types of equations. Some have rectilinear (i.e, straight rather than curved) scales for all three variables, but these scales can lie at angles relative to each other or may even radiate from a single point. A nomogram can also consist of a mix of curved and rectilinear scales. For an equation of more than three variables, two of the variables can form a grid of curves for their values, in which an intersection point within the grid serves as one of the crossing points for the isopleth. However, Gronwall's paper is limited to three-variable equations represented by three scales, to which we will limit our discussion.

Some of the methods of laying out and marking the scales of a nomogram are geometric in nature, but the approach having the greatest success in creating and transforming nomograms involves determinant equations [**267**]. This is the starting point of Gronwall's paper, and so a brief review is in order.

Consider a three-variable equation in x, y and z,
$$(1) \qquad F(x, y, z) = 0.$$

As Gronwall states at the beginning of his paper, a nomogram can be constructed for this equation if it can be represented in the form of a determinant equation of the form

(2) $$\begin{vmatrix} f_1(x) & g_1(x) & h_1(x) \\ f_2(y) & g_2(y) & h_2(y) \\ f_3(z) & g_3(z) & h_3(z) \end{vmatrix} = 0,$$

where each variable is constrained to a separate row in the determinant.

If the functions $h_1(x)$, $h_2(y)$ and $h_3(z)$ are non-zero (or as Gronwall puts it, "not identically canceled out"), we can divide each row by its third element and obtain new values of f_i and g_i in what is called the standard determinant form of the equation:

(4) $$\begin{vmatrix} f_1(x) & g_1(x) & 1 \\ f_2(y) & g_2(y) & 1 \\ f_3(z) & g_3(z) & 1 \end{vmatrix} = 0.$$

Expanding the determinant according to the rules of linear algebra gives the equation

$$f_1(x)g_2(y) + g_1(x)f_3(z) + f_2(y)g_3(z) - g_2(y)f_3(z) - g_1(x)f_2(y) - f_1(x)g_3(z) = 0,$$

and this equation must be equivalent to the original equation of the nomogram.

Once we have expressed an equation as a determinant equation in standard form, drawing a nomogram for it is straightforward. Each row represents a scale for a variable, and f_i and g_i are the Cartesian coordinates of the values of that variable. Gronwall assigns ξ and η as the Cartesian coordinates to distinguish them from the variables x and y in the equation.[1] For the scale of x, for example, functions $f_1(x) = 0$ and $g_1(x) = 2x$ mean that (ξ, η) is $(0, 2x)$ for any value x, or in other words, the x scale is vertical at $\xi = 0$ and positive and negative values of x are marked at locations $2x$ along the η-axis. Functions $f_1(x) = x$ and $g_1(x) = x^2$ imply $(\xi, \eta) = (x, x^2)$, so the value $x = 2$ is marked at $(2, 4)$ and so forth. In this role, f_i and g_i represent parametric functions of the scale, and since $g_1(x) = f_1^2(x)$ in the latter example, the scale will lie along a parabolic curve.

By expanding the determinant, we can verify that the determinant equation satisfies the original equation, but it is sometimes very difficult or impossible to find the appropriate elements f_i and g_i of the determinant. There is a way of creating an initial determinant equation from the equation, but one that is not yet in standard form, i.e., the variables are not isolated into individual rows and the third column is not reduced to ones.

---

[1] To avoid this awkwardness, other texts use variables such as u, v and w for the variables in the equation.

3.1. THE DESIGN OF NOMOGRAMS

Consider the equation
$$z = x + y, \quad \text{or} \quad z - x - y = 0.$$

Let's define $A = x$ and $B = y$. Then we have the system of equations,
$$\begin{cases} 1 \times A + 0 \times B - x = 0, \\ 0 \times A + 1 \times B - y = 0, \\ -1 \times A - 1 \times B + z = 0, \end{cases}$$
where the first two equations are simply re-arrangements of the definitions of A and B above, and the third equation is from our original equation $z - x - y = 0$. This system of equations in A and B implies a determinant equation containing the coefficients of each term,

(3.1.1)
$$\begin{vmatrix} 1 & 0 & -x \\ 0 & 1 & -y \\ -1 & -1 & z \end{vmatrix} = 0.$$

If we can convert this determinant equation into standard form, we have designed a nomogram for the equation $z - x - y = 0$. The following operations are allowed when a determinant such as this is equal to zero:

- Rows and columns can be swapped.
- Rows and columns can be multiplied or divided by any non-zero number or function.
- A multiple of the elements in any row or column can be added to their corresponding elements in any other row or column.

Returning to our determinant equation (3.1.1), we can perform the following operations:

(1) Add column 2 to column 1;
(2) Divide row 3 by -2;
(3) Multiply column 3 by -1; and
(4) Move column 1 to be column 3.

We end up with the standard determinant equation,

(3.1.2)
$$\begin{vmatrix} 0 & x & 1 \\ 1 & y & 1 \\ \frac{1}{2} & \frac{z}{2} & 1 \end{vmatrix} = 0.$$

Here we see that the x scale is vertical at $\xi = 0$ and each value of x is marked at $\eta = x$. The y scale is vertical at $\xi = 1$ and marked likewise. The

z scale is vertical at $\xi = \frac{1}{2}$ and each value of z is marked at $\eta = \frac{z}{2}$. This nomogram is shown in Figure 3.1.

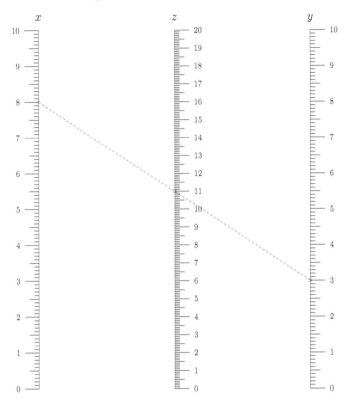

FIGURE 3.1. Nomogram for $z = x + y$.

Note that we can replace the variables x, y and z by functions of these variables. Also, equations consisting of products of powers of functions can be converted to this form by the use of logarithms, as $z = x^2 y$ can be written as $\log z = 2 \log x + \log y$. Each term is then substituted into the determinant equation (3.1.2). This versatility makes the parallel scale nomogram for addition the most commonly encountered type of nomogram.

For example, the equation

(3.1.3) $$z = 5x + 10y + 2xy$$

can be converted to the form $z + 25 = (2x + 10)(y + 2.5)$ or $\log(z + 25) = \log(2x + 10) + \log(y + 2.5)$. When these logarithmic terms are substituted into the standard determinant equation (3.1.2) we have the parallel scale nomogram depicted in Figure 3.2.

3.1. THE DESIGN OF NOMOGRAMS

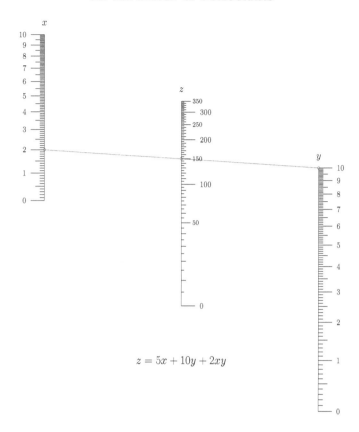

FIGURE 3.2. Nomogram for $z = 5x + 10y + 2xy$.

Now since the determinant equals zero, we can multiply it by any other determinant and the determinant equation will still hold. The general form of such a multiplier is

(3.1.4)
$$\begin{vmatrix} a_1 & a_2 & a_3 \\ b_1 & b_2 & b_3 \\ c_1 & c_2 & c_3 \end{vmatrix},$$

and we use Gronwall's notation for the new elements of the standard determinant equation after multiplication by this new determinant,

(5)
$$\begin{vmatrix} \overline{f_1}(x) & \overline{g_1}(x) & 1 \\ \overline{f_2}(y) & \overline{g_2}(y) & 1 \\ \overline{f_3}(z) & \overline{g_3}(z) & 1 \end{vmatrix} = 0,$$

where

(6)
$$\begin{cases} \overline{f}_i = \dfrac{a_1 f_i + b_1 g_i + c_1}{a_3 f_i + b_3 g_i + c_3}, \\ \overline{g}_i = \dfrac{a_2 f_i + b_2 g_i + c_2}{a_3 f_i + b_3 g_i + c_3} \end{cases} \quad (i = 1, 2, 3).$$

These are called *homographic transformations*, and Gronwall defines a *family* of nomograms as those that are related by a homographic tranformation. In his paper Gronwall derives unique families of nomograms with the understanding that each of these has a myriad of forms related by these homographic transformations.

These transformations can make radical changes in the appearance of nomograms. In the simplest case, a homographic transformation can be used to square up a nomogram to better fit a sheet of paper. Here the values a_i, b_i and c_i are found by solving the system of equations obtained by equating the maximum values of $\overline{f}_i$ and $\overline{g}_i$ to the corner coordinates.[**258**, pp. 56–57] This extends the lengths of scales for greater precision (and shifts the middle scale in the process), as shown in Figure 3.3.

We can also perform standard projective transformations on the nomogram with homographic transformations (6) as

(3.1.5)
$$\begin{cases} \overline{f}_i = f_i + m, \quad \overline{g}_i = g_i + n & \text{translation} \\ \overline{f}_i = f_i \sin\theta + g_i \cos\theta, \quad \overline{g}_i = f_i \cos\theta - g_i \sin\theta & \text{rotation} \\ \overline{f}_i = m f_i, \quad \overline{g}_i = n g_i & \text{stretch} \\ \overline{f}_i = f_i \cos\theta, \quad \overline{g}_i = g_i + f_i \sin\theta & \text{shear} \\ \overline{f}_i = \dfrac{z_P f_i}{f_i - x_P}, \quad \overline{g}_i = \dfrac{y_P f_i - x_P g_i}{f_i - x_P} & \text{projection via } P(x_p, y_p, z_p) \end{cases}$$

But there are many other forms of nomogram we can draw for an equation, each corresponding to a different standard determinant equation (4) that yields the same equation when expanded. For example, our equation $z = 5x + 10y + 2xy$ or $z + 25 = (2x + 10)(y + 2.5)$ can also be written as

(3.1.6)
$$\begin{vmatrix} \dfrac{1}{1+(2x+10)^2} & \dfrac{2x+10}{1+(2x+10)^2} & 1 \\ \dfrac{1}{1+(y+2.5)^2} & -\dfrac{y+2.5}{1+(y+2.5)^2} & 1 \\ \dfrac{1}{1+(z+25)} & 0 & 1 \end{vmatrix} = 0.$$

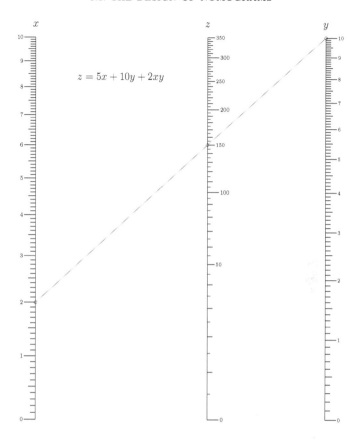

FIGURE 3.3. Nomogram with rectangular outline for $z = 5x + 10y + 2xy$.

This results in the circular nomogram of Figure 3.4. Note that the entire range of each variable is now contained in a scale of finite length. In contrast to some of the other nomograms in this chapter, circular nomograms have been in common use for many years in practical applications [253][271][257].

We find in Figure 3.4 that the sample isopleth for our region of interest around $x = 2, y = 10$ lies in a compressed area at the very left side of the circle. However, we can multiply the factors $2x + 10$ and $y + 2.5$ in our equation by constants m and n if we also multiply $z + 25$ by mn. This allows us to shift the x and y scales independently to bring our region of interest into the most precise central area as shown in Figure 3.5. This is an example of the use of *parameters* to vary the properties of a family of nomograms. For our earlier equation $z = x + y$, we can multiply each variable by a constant, and we can also shift ranges if we replace the functions x, y and z with $x - x_0$, $y - y_0$ and

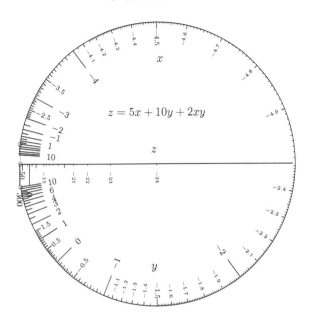

FIGURE 3.4. Circular nomogram for $z = 5x + 10y + 2xy$.

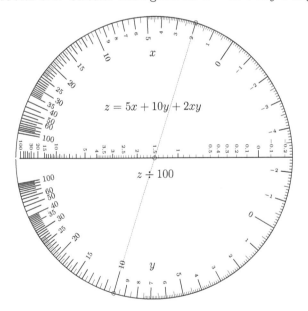

FIGURE 3.5. Circular nomogram with optimized ranges for $z = 5x + 10y + 2xy$.

$z - z_0$ with $z_0 = x_0 + y_0$. Gronwall defines a *class* of equations as those related by such parameters, and he is careful to list the free parameters for the various families of nomograms he derives in Section 4 of his paper for the equation $x + y + z = 0$, or more generally, $\phi(x) + \psi(y) + \chi(z) = 0$.

The point here is that nomograms are malleable—very malleable—once their equations are put into the form of a determinant equation. In his paper, Gronwall is interested in identifying independent families of nomograms not related by homographic transformations. However, we should be aware that each of these families is in fact composed of a wide variety of interesting variations.

So the first step in creating a nomogram is to convert the equation into a standard determinant equation. Then we can plot each scale of the nomogram from the first two elements in each row of the determinant. Finally, we can perform homographic transformations on the determinant and vary the free parameters of the equation to morph the nomogram into a configuration optimum for our use. Gronwall describes this background information on nomograms in a cursory manner at the beginning of his paper, to which we now proceed.

3.2. The Goals and Structure of Gronwall's Paper

Gronwall's primary goal in his paper is stated quite simply at the beginning: to find a way of determining whether a three-variable equation $F(x, y, z) = 0$ can be represented by a nomogram, or more precisely, that it can be cast as a determinant equation in the form

(4)
$$\begin{vmatrix} f_1(x) & g_1(x) & 1 \\ f_2(y) & g_2(y) & 1 \\ f_3(z) & g_3(z) & 1 \end{vmatrix} = 0.$$

Gronwall accomplishes this relatively early in his paper, at least to the point he was able to take his analysis. Gronwall found that a necessary and sufficient condition for the existence of such a nomogram is that the system of two partial differential equations in (24) has a common integral C (i.e, a function that satisfies both equations).

(24)
$$\begin{cases} M\dfrac{\partial^2 C}{\partial x^2} + \dfrac{\partial^2 C}{\partial x \partial y} = \left(MC - 2\dfrac{\partial M}{\partial x}\right)\dfrac{\partial C}{\partial x} + 2C\dfrac{\partial C}{\partial y} \\ \qquad + \dfrac{\partial M}{\partial x}C^2 + \left(\dfrac{\partial N}{\partial x} - \dfrac{\partial^2 M}{\partial x^2}\right)C - \dfrac{\partial^2 N}{\partial x^2}, \\ 2M\dfrac{\partial^2 C}{\partial x \partial y} + \dfrac{\partial^2 C}{\partial y^2} = 2\left(M^2 C + MN - \dfrac{\partial M}{\partial y}\right)\dfrac{\partial C}{\partial x} \\ \qquad + \left(MC + N - 2\dfrac{\partial M}{\partial x}\right)\dfrac{\partial C}{\partial y} \\ \qquad + 2M\dfrac{\partial M}{\partial x}C^2 + 2\left(N\dfrac{\partial N}{\partial x} + M\dfrac{\partial N}{\partial x} - \dfrac{\partial^2 M}{\partial x \partial y}\right)C \\ \qquad + 2N\dfrac{\partial N}{\partial x} - 2\dfrac{\partial^2 N}{\partial x \partial y}. \end{cases}$$

The functions M and N are defined as

(14)
$$\begin{cases} M = -\dfrac{\frac{\partial z}{\partial y}}{\frac{\partial z}{\partial x}}, \\ N = \dfrac{\partial M}{\partial x} + \dfrac{1}{M}\dfrac{\partial M}{\partial y} = \dfrac{\left(\frac{\partial z}{\partial y}\right)^2 \frac{\partial^2 z}{\partial x^2} - 2\frac{\partial z}{\partial x}\frac{\partial z}{\partial y}\frac{\partial^2 z}{\partial x \partial y} + \left(\frac{\partial z}{\partial x}\right)^2 \frac{\partial^2 z}{\partial y^2}}{\left(\frac{\partial z}{\partial x}\right)^2 \frac{\partial z}{\partial y}}. \end{cases}$$

Gronwall derives a number of other auxiliary functions as well, the most important being D, where

(18) $$D = MC + N.$$

In terms of D, the system of equations (24) is written as

(23)
$$\begin{cases} 2\dfrac{\partial^2 C}{\partial x \partial y} + \dfrac{\partial^2 D}{\partial x^2} - C\left(2\dfrac{\partial C}{\partial y} + \dfrac{\partial D}{\partial x}\right) = 0, \\ \dfrac{\partial^2 C}{\partial y^2} + 2\dfrac{\partial^2 D}{\partial x \partial y} - D\left(\dfrac{\partial C}{\partial y} + 2\dfrac{\partial D}{\partial x}\right) = 0. \end{cases}$$

In addition, Gronwall claims that "essentially distinct nomographic representations" occur only when

(50) $$\dfrac{\partial^2 \log M}{\partial x \partial y} = 0.$$

We will see that this is the necessary and sufficient condition for the case where all three scales of a nomogram can be rectilinear. Gronwall means here that there are different families of nomogram for this case that cannot be transformed into each other through a homographic transformation as defined in (6) and described in Section 3.1.

3.2. THE GOALS AND STRUCTURE OF GRONWALL'S PAPER

These results are mathematically sound, but if the paper had ended at that point it would have been a paper of even more limited academic interest. It is an advancement over the work of Duporcq he cites, in which only some sufficient conditions were presented. However, the system of equations certainly appears prohibitively difficult to solve. Gronwall implicitly acknowledges this in his paper when he offers to construct an explicit solution to (24) in a later paper, a formidable goal that proved elusive. What makes this paper such an important work in the history of nomography is that Gronwall, with his innate focus on finding practical solutions, exploits properties of (50) and (24) to analyze different forms of equations and nomograms.

In the process Gronwall lays out procedures for finding the elements in the standard determinant equation for different forms of nomogram. These includes nomograms with one curved and two rectilinear scales and nomograms with one rectilinear and two curved scales. In his final section he considers the special case where the two curved scales in the latter case are coincident along a conic, a type of nomogram discovered by J. Clark. Gronwall also provides individual tests to determine whether the scale for each variable x, y and z can be represented by a rectilinear scale, although these generally require knowledge of C.

The most fascinating discussion, though, appears in Section 4 of the paper, a detailed treatment of equations that meet the test for three rectilinear scales given in (50). This test does not require C from (24), so it is relatively easy to perform on an equation we are trying to cast as a nomogram. If our equation passes the test, it can be written as a simple sum of functions of the variables, as in $\phi(x) + \psi(y) + \chi(z) = 0$. Once it passes the test, Gronwall provides the means to convert the equation into this general form, and as we have seen in Section 3.1 it is then a straightforward process to create a nomogram of three parallel scales.

But Gronwall has told us that there are other nomographic representations for this case, and he proceeds to derive every one of them. He provides determinant equations for several families of nomogram, each associated with a different C, as well as free parameters associated with different classes of equations within the family. These parameters can be used, for example, to shift ranges of interest for x, y and z into areas of the nomogram offering the most precision. Some of these cases result in nomograms in which two of the scales share a conic, another nod to Clark. Remarkably, Gronwall also derives cases in which all three scales share the same curve, one that is cut across in three places by an isopleth, thereby again extending work done by Clark to additional families of nomograms.

In the end, Gronwall succeeds for the first time in providing a new and complete criterion, complicated as it is, for an equation that can be represented by a nomogram. He provides a more robust theoretical basis for general three-variable nomograms, including Clark's conical and single-curve nomograms. And he discovers new forms of nomograms and catalogues all possible families and classes of nomograms for equations that satisfy the criterion (50) for three rectilinear scales.

The criterion (24) for equations that can represented by a nomogram is presented in Section 1 of the paper. A set of auxiliary functions is also defined in this section, and we will require those to ultimately find the elements of the standard determinant equation when (50) does not hold. Section 2 of the paper derives tests for determining whether individual scales are rectilinear or curved, including the criterion (50) for equations whose nomogram can consist of three rectilinear scales. Section 3 details the steps necessary to find the determinant elements for nomograms with one rectilinear and two curved scales. Section 4 covers equations that meet (50) that we discussed above. Section 5 details a new method for finding the determinant elements for nomograms with one curved and two rectilinear scales. Section 6 treats Clark's conical nomograms. We will follow this narrative fairly closely, with the exception that the method for converting an equation into the sum of simple functions at the start of Gronwall's Section 4 is moved up with his discussion of that criterion (50) at the end of his Section 2.

3.3. Tests for Rectilinear Scales

In his Section 2, Gronwall proceeds to derive tests that can be performed on an equation to determine whether any or all of the scales of its nomogram can be rectilinear. We can always perform operations on the determinant equation to force a scale to be curved, but in general the nomograms with the most rectilinear scales are the easiest to draw and to use. Also, there are other tests of curved-scale nomograms (such as the conical nomograms covered later by Gronwall) that are not usable if certain scales are rectilinear. Therefore, a reasonable first step in analyzing an equation is to determine which of the scales can be rectilinear.

First, Gronwall derives a test for the x scale to be rectilinear as

(37) $$\frac{\partial C}{\partial y} + 2\frac{\partial D}{\partial x} = 0.$$

A similar test is provided for the y scale,

(45) $$2\frac{\partial C}{\partial y} + \frac{\partial D}{\partial x} = 0,$$

and for the z scale,

(47) $$\frac{\partial C}{\partial y} - \frac{\partial D}{\partial x} + 3\frac{\partial^2 \log M}{\partial x \partial y} = 0.$$

A test for both the x and y scales to be rectilinear is given as

(48) $$\frac{\partial C}{\partial y} = \frac{\partial D}{\partial x} = 0.$$

There is asymmetry among the three scales in the equations above despite the symmetry evident in the general determinant (4). This is due to Gronwall's initial decision in (7) to define u and v in terms of the third row (in z), as well as the roles of x and y in the definition of M in (14). The x-y asymmetry is apparent in Gronwall's comparison of the corresponding functions above (45) in his paper. We will be using these tests in later sections to determine the configuration of scales for different equations, since Gronwall has simplified his procedures for different cases.

Now a very important test is for the case where all three scales can be rectilinear. The necessary and sufficient condition for this situation is

(50) $$\frac{\partial^2 \log M}{\partial x \partial y} = 0.$$

Gronwall devotes a lengthy section to the implications of passing this test, which corresponds to an equation that can be written as

(51) $$\phi(x) + \psi(y) + \chi(z) = 0.$$

In his Section 4, treated later in this chapter, Gronwall derives all possible classes of nomograms for this equation. Again, the scales do not necessarily have to be drawn as rectilinear. In fact, we will see that most of the nomograms he catalogues in this instance do not have three rectilinear scales, and in some cases none of the scales are rectilinear.

Unlike the tests for individual or paired rectilinear scales, there is no need to find C to determine if all three scales can be rectilinear. Therefore, it is a good idea to perform this test if there might be some question about it. It is not always obvious that an equation can be written as a sum of individual functions of x, y and z.

Once an equation is written in the form $\phi(x) + \psi(y) + \chi(z) = 0$, we can write the standard determinant equation as Gronwall does in his equation below

(51):

(3.3.1)
$$\begin{vmatrix} \phi(x) & -1 & 1 \\ \psi(y) & 1 & 1 \\ -\frac{1}{2}\chi(z) & 0 & 1 \end{vmatrix} = 0.$$

The nomogram corresponding to this determinant appears in Figure 3.6. This is standard parallel scale nomogram oriented horizontally. It is compressed vertically by a factor of two here.

FIGURE 3.6. Nomogram for $\phi(x) + \psi(y) + \chi(z) = 0$.

Let's consider the equation $z = ax + by + cxy + d$. Can we draw this nomogram as three parallel scales in functions of x, y and z?

First, we find M from (14) and perform the test in (50), giving

$$M = -\frac{\frac{\partial z}{\partial y}}{\frac{\partial z}{\partial x}} = -\frac{b+cx}{a+cy},$$

$$\log M = \log(a+cy) - \log(b+cx),$$

$$\frac{\partial^2 \log M}{\partial x \partial y} = 0.$$

We now know that this equation can be written as the sum of functions of x, y and z. We can also deduce this new form of the equation using a method Gronwall describes at the start of his Section 4. If (50) holds, then

(65) $$M = \alpha(x)\beta(y),$$

and given (51) we find that

(66) $$\begin{cases} \phi(x) = \int \dfrac{dx}{\alpha(x)}, \\ \psi(y) = -\int \beta(y)dy. \end{cases}$$

Here M is guaranteed to be the simple product of a function of x and a function of y. In this case our value of M allows us to write $\alpha(x) = b + cx$ and $\beta(y) = -(a + cy)$. Then

$$\begin{cases} \phi(x) = \int \dfrac{dx}{b + cx} = \dfrac{1}{c}\log(b + cx), \\ \psi(y) = -\int \dfrac{-dy}{a + cy} = \dfrac{1}{c}\log(a + cy). \end{cases}$$

We can find $\chi(z)$ from the overall relation $\phi(x) + \psi(y) + \chi(z) = 0$. Since we met the test (50), we are guaranteed that we can express $\chi(z)$ in terms of z alone. Constants of integration also cancel out here.

$$\chi(z) = -\dfrac{1}{c}\log(b + cx) - \dfrac{1}{c}\log(a + cy)$$
$$= -\dfrac{1}{c}\log(ab + acx + bcy + c^2xy).$$

But from the original equation we have $z = ax + by + cxy + d$, so

$$\chi(z) = -\dfrac{1}{c}\log(cz - cd + ab)$$

Now we put these functions into the general equation $\phi(x) + \psi(y) + \chi(z) = 0$ and simplify it. The resulting equation is

(3.3.2) $$\log(b + cx) + \log(a + cy) - \log(cz - cd + ab) = 0,$$

which is equivalent to factoring $z = ax + by + cxy + d$ as $z - d + \frac{ab}{c} = (cx + b)(y + \frac{a}{c})$. Comparing our (3.3.2) with the determinant equation (3.3.1) for $\phi(x) + \psi(y) + \chi(z) = 0$, and swapping the first two columns so the nomogram is in the traditional vertical orientation, we find the determinant equation

$$\begin{vmatrix} -1 & \log(b + cx) & 1 \\ 1 & \log(a + cy) & 1 \\ 0 & \dfrac{\log(cz - cd + ab)}{2} & 1 \end{vmatrix} = 0.$$

Figure 3.7 shows this nomogram for $a = 2$, $b = 4$, $c = 1$ and $d = 5$.

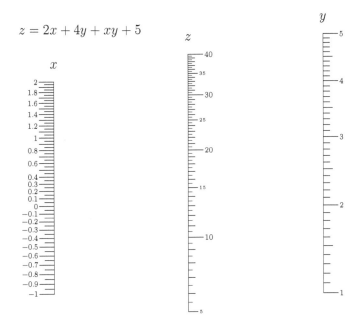

FIGURE 3.7. Nomogram for $z = 2x + 4y + xy + 5$.

This is essentially equivalent to a method put forth by Paul de Saint-Robert in 1867 as the criterion for an equation to be represented by two fixed scales and a sliding scale, such as a special slide rule [**259**, pp. 301–309][**269**, pp. 17–21].

Factoring our equation by hand rather than this method may not seem very difficult, but there are more complicated equations for which this method is useful. For example, it is not obvious at all that the equation

$$z = xy + \sqrt{1+x^2}\sqrt{1+y^2}$$

can be written as

$$\log(z + \sqrt{z^2 - 1}) = \log(x + \sqrt{1+x^2}) + \log(y + \sqrt{1+y^2}),$$

but this is found directly from the method outlined by Gronwall.

3.4. The Case of One Rectilinear and Two Curved Scales

In Section 3 of his paper, Gronwall surveys the most difficult case of all the ones he considers, a nomogram in which one of the scales is rectilinear and the other two scales are curved, as in Figure 3.8. The rectilinear scale can be located inside or outside the curved scales.

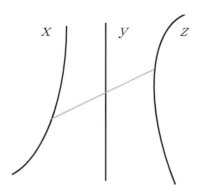

FIGURE 3.8. Nomogram with two curved scales and one rectilinear scale.

Here he assumes that the common solution C of the system of differential equations in (24) has been found for the equation of the nomogram. For the simpler cases in the later sections, direct formulas for C are provided. Gronwall also assumes here that the scales of x and y are curved and the scale for z is rectilinear; he remarks that the method is general enough when the scale variables are otherwise, but in fact a simple change of variables is all that is needed to convert it to the assumed form.

To explore the method described in this section, let's create a nomogram for the equation

(3.4.1) $$z = \frac{xy}{x+y},$$

which is equivalent to the harmonic relation $\frac{1}{z} = \frac{1}{x} + \frac{1}{y}$.

We are trying to find the elements in the standard determinant form of this equation, or

(4) $$\begin{vmatrix} f_1(x) & g_1(x) & 1 \\ f_2(y) & g_2(y) & 1 \\ f_3(z) & g_3(z) & 1 \end{vmatrix} = 0.$$

From (14),

$$\begin{cases} M = -\dfrac{\frac{\partial z}{\partial y}}{\frac{\partial z}{\partial x}} = -\dfrac{\frac{x^2}{(x+y)^2}}{\frac{y^2}{(x+y)^2}} \\ \quad = -\dfrac{x^2}{y^2}, \\ N = \dfrac{\partial M}{\partial x} + \dfrac{1}{M}\dfrac{\partial M}{\partial y} = \dfrac{-2x}{y^2} + \dfrac{-y^2}{x^2}\dfrac{2x^2}{y^3} \\ \quad = -\dfrac{2(x+y)}{y^2}. \end{cases}$$

Now we substitute our values M and N into (24) to arrive at the system of equations for C:

$$\begin{cases} -\dfrac{x^2}{y^2}\dfrac{\partial^2 C}{\partial x^2} + 2\dfrac{\partial^2 C}{\partial x \partial y} = \left(-\dfrac{x^2}{y^2}C - 2\dfrac{-2x}{y^2}\right)\dfrac{\partial C}{\partial x} + 2C\dfrac{\partial C}{\partial y} \\ \qquad + \dfrac{-2x}{y^2}C^2 + \left(\dfrac{-2}{y^2} - \dfrac{-2}{y^2}\right)C - 0, \\ 2\dfrac{-x^2}{y^2}\dfrac{\partial^2 C}{\partial x \partial y} + \dfrac{\partial^2 C}{\partial y^2} = 2\left(\dfrac{x^4}{y^4}C + \dfrac{-x^2}{y^2}\dfrac{-2(x+y)}{y^2} - \dfrac{2x}{y^3}\right)\dfrac{\partial C}{\partial x} \\ \qquad + \left(-\dfrac{x^2}{y^2}C + \dfrac{-2(x+y)}{y^2} - 2\dfrac{-2x}{y^2}\right)\dfrac{\partial C}{\partial y} + 2\dfrac{-x^2}{y^2}\dfrac{-2x}{y^2}C^2 \\ \qquad + 2\left(\dfrac{-2(x+y)}{y^2}\dfrac{-2x}{y^2} + \dfrac{-x^2}{y^2}\dfrac{-2}{y^2} - \dfrac{4x}{y^3}\right)C \\ \qquad + 2\dfrac{-2(x+y)}{y^2}\dfrac{-2}{y^2} - 2\dfrac{4}{y^3}, \end{cases}$$

or, simplified,

$$\begin{cases} x^2\dfrac{\partial^2 C}{\partial x^2} - 2y^2\dfrac{\partial^2 C}{\partial x \partial y} = (x^2 C - 4x)\dfrac{\partial C}{\partial x} - 2y^2 C\dfrac{\partial C}{\partial y} + 2xC^2, \\ -2x^2 y^2 \dfrac{\partial^2 C}{\partial x \partial y} + y^4 \dfrac{\partial^2 C}{\partial y^2} = 2(x^4 C + 2x^2(x+y) - 2xy)\dfrac{\partial C}{\partial x} + \\ \qquad (-x^2 y^2 C + 2xy^2 - 2y^3)\dfrac{\partial C}{\partial y} + 4x^3 C^2 + 12x^2 C + 8x. \end{cases}$$

These are not straightforward to solve, of course, even with the symbolic mathematics available in computer algebra systems. Here we posit a solution

(3.4.2) $$C = \dfrac{2}{y-x}$$

and simply verify that it satisfies the two equations. Since we have C, it is easier to find $D = MC + N$ from (18) and use the simpler versions (23) of

3.4. THE CASE OF ONE RECTILINEAR AND TWO CURVED SCALES

Gronwall's equations:

(23)
$$\begin{cases} 2\dfrac{\partial^2 C}{\partial x \partial y} + \dfrac{\partial^2 D}{\partial x^2} - C\left(2\dfrac{\partial C}{\partial y} + \dfrac{\partial D}{\partial x}\right) = 0, \\ \dfrac{\partial^2 C}{\partial y^2} + 2\dfrac{\partial^2 D}{\partial x \partial y} - D\left(\dfrac{\partial C}{\partial y} + 2\dfrac{\partial D}{\partial x}\right) = 0 \end{cases}$$

Gronwall had derived (24) from (23) by replacing D with $MC + N$. Now

$$D = MC + N$$
$$= \frac{-x^2}{y^2} \frac{2}{y - x} - \frac{2(x + y)}{y^2}$$
$$= -\frac{2}{y - x}$$

so

$$\begin{cases} 2\dfrac{-4}{(y-x)^3} + \dfrac{-4}{(y-x)^3} - \dfrac{2}{y-x}\left(2\dfrac{-2}{(y-x)^2} + \dfrac{-2}{(y-x)^2}\right) \\ \quad = \dfrac{-8}{(y-x)^3} + \dfrac{-4}{(y-x)^3} + \dfrac{8}{(y-x)^3} + \dfrac{4}{(y-x)^3} \\ \quad = 0, \\ \dfrac{4}{(y-x)^3} + 2\dfrac{4}{(y-x)^3} - \dfrac{-2}{y-x}\left(\dfrac{-2}{(y-x)^2} + 2\dfrac{-2}{(y-x)^2}\right) \\ \quad = \dfrac{4}{(y-x)^3} + \dfrac{8}{(y-x)^3} - \dfrac{4}{(y-x)^3} - \dfrac{8}{(y-x)^3} \\ \quad = 0, \end{cases}$$

and we have verified that $C = \frac{2}{y-x}$ is a solution.

What form of nomogram do we expect? We can use the tests in Section 3.3 to find out whether the scales are curved or rectilinear.

The necessary and sufficient condition for the x scale to be rectilinear is given by

(37)
$$\frac{\partial C}{\partial y} + 2\frac{\partial D}{\partial x} = 0,$$

and we find the x scale is curved because

$$\frac{-2}{(y-x)^2} + 2\frac{-2}{(y-x)^2} \neq 0.$$

The necessary and sufficient condition for the y scale to be rectilinear is

(45)
$$2\frac{\partial C}{\partial y} + \frac{\partial D}{\partial x} = 0,$$

and we find the y scale is curved because
$$2\frac{-2}{(y-x)^2} + \frac{-2}{(y-x)^2} \neq 0.$$

The necessary and sufficient condition for the z scale to be rectilinear is

(47) $$\frac{\partial C}{\partial y} - \frac{\partial D}{\partial x} + 3\frac{\partial^2 \log M}{\partial x \partial y} = 0,$$

and we find the z scale is rectilinear because
$$\frac{-2}{(y-x)^2} - \frac{-2}{(y-x)^2} + 3(0) = 0.$$

We now know that the nomogram can be represented by curved x and y scales and a rectangular z scale. To design it, we will follow Gronwall's procedure outlined in the penultimate paragraph of his Section 3, and in more detail in the second and third paragraphs under equation (35) in his Section 1, to derive the elements f_i and g_i of the standard determinant equation (4). This procedure works only with curved x and y scales because, as we will see, (37) and (45) appear in a denominator of one of the intermediate formulas.

In his Section 1, Gronwall arrives at a linear system of equations in ω that must be satisfied for a common solution (or integral) of (24):

(26) $$\begin{cases} \dfrac{\partial^2 \omega}{\partial x^2} = \dfrac{1}{3}C\dfrac{\partial \omega}{\partial x} + \dfrac{1}{3}\left(\dfrac{2}{3}C^2 - \dfrac{\partial C}{\partial x}\right)\omega, \\[6pt] \dfrac{\partial^2 \omega}{\partial x \partial y} = -\dfrac{1}{3}D\dfrac{\partial \omega}{\partial x} - \dfrac{1}{3}C\dfrac{\partial \omega}{\partial y} + \dfrac{1}{3}\left(-\dfrac{1}{3}CD + \dfrac{\partial C}{\partial y} + \dfrac{\partial D}{\partial x}\right)\omega, \\[6pt] \dfrac{\partial^2 \omega}{\partial y^2} = \dfrac{1}{3}D\dfrac{\partial \omega}{\partial y} + \dfrac{1}{3}\left(\dfrac{2}{3}D^2 - \dfrac{\partial D}{\partial y}\right)\omega. \end{cases}$$

This is a linear system of equations, that is, any derivative of ω is a linear function of ω, $\frac{\partial \omega}{\partial x}$ and $\frac{\partial \omega}{\partial y}$. This means that the solution ω can be expressed as the sum of three particular integrals ω_1, ω_2 and ω_3, or

(3.4.3) $$\omega = \alpha\omega_1 + \beta\omega_2 + \gamma\omega_3.$$

A fundamental system ω_1, ω_2, ω_3 requires that the following determinant equation holds for a non-zero constant:

(28) $$\Delta(\omega_1, \omega_2, \omega_3) = \begin{vmatrix} \omega_1 & \frac{\partial \omega_1}{\partial x} & \frac{\partial \omega_1}{\partial y} \\ \omega_2 & \frac{\partial \omega_2}{\partial x} & \frac{\partial \omega_2}{\partial y} \\ \omega_3 & \frac{\partial \omega_3}{\partial x} & \frac{\partial \omega_3}{\partial y} \end{vmatrix} = const. \neq 0.$$

3.4. THE CASE OF ONE RECTILINEAR AND TWO CURVED SCALES

If a set of functions w_1, w_2 and w_3 can be found that form such a fundamental system of (26), then we can determine functions u and v of Gronwall's Section 1 from

(30)
$$\begin{cases} u = \dfrac{w_1}{w_3}, \\ v = \dfrac{w_2}{w_3}. \end{cases}$$

Then we can find f_1 and f_2 from

(11)
$$\begin{cases} f_1(x) = -\dfrac{\frac{\partial v}{\partial y}}{\frac{\partial u}{\partial y}}, \\ f_2(y) = -\dfrac{\frac{\partial v}{\partial x}}{\frac{\partial u}{\partial x}}, \end{cases}$$

and g_1 and g_2 from

(9)
$$\begin{cases} g_1(x) = u f_1(x) + v, \\ g_2(y) = u f_2(y) + v, \end{cases}$$

and f_3 and g_3 from

(15)
$$f_3(z) = -\dfrac{M \frac{\partial v}{\partial x} + \frac{\partial v}{\partial y}}{M \frac{\partial u}{\partial x} + \frac{\partial u}{\partial y}}$$

and

(7)
$$g_3(z) = u f_3(z) + v.$$

Gronwall shows that f_1 and g_1 will necessarily turn out to be functions of x only, f_2 and g_2 of y only, and f_3 and g_3 of z only.

So in his Section 3, Gronwall provides a way of finding the fundamental system w_1, w_2, w_3, and it all cascades from there to the standard determinant equation (4) from which the nomogram can be constructed. The development leads to an equation for w_i, where $i = 1, 2, 3$. Gronwall shows that if scales x and y are curved, we can select constants x_1, x_2, x_3 in the following equation (60) such that w_1, w_2, w_3 form a fundamental system of (26).

(60)
$$w_i = \dfrac{1}{\left(2\frac{\partial C}{\partial y} + \frac{\partial D}{\partial x}\right)_{x=x_i}^{\frac{2}{3}} \left(\frac{\partial C}{\partial y} + 2\frac{\partial D}{\partial x}\right)_{x=x_i}^{\frac{1}{3}} \left(2\frac{\partial C}{\partial y} + \frac{\partial D}{\partial x}\right)^{\frac{1}{3}}}$$
$$\times \left\{ 2\frac{\partial}{\partial y} \log\left(2\frac{\partial C}{\partial y} + \frac{\partial D}{\partial x}\right) + D \right.$$
$$\left. - \left[2\frac{\partial}{\partial y} \log\left(2\frac{\partial C}{\partial y} + \frac{\partial D}{\partial x}\right) + D\right]_{x=x_i} \right\}, \quad (i=1,2,3).$$

Here is where we see (37) and (45) in the denominator, requiring the x and y scales to be curved.

Resuming our work on the equation $z = \frac{xy}{x+y}$, we find three of the terms in (60) from our C and D as

$$\frac{\partial C}{\partial y} + 2\frac{\partial D}{\partial x} = \frac{-2}{(y-x)^2} + 2\frac{-2}{(y-x)^2} = -\frac{6}{(y-x)^2}$$

$$2\frac{\partial C}{\partial y} + \frac{\partial D}{\partial x} = 2\frac{-2}{(y-x)^2} + \frac{-2}{(y-x)^2} = -\frac{6}{(y-x)^2}$$

$$2\frac{\partial}{\partial y}\log\left(2\frac{\partial C}{\partial y} + \frac{\partial D}{\partial x}\right) + D = 2\frac{\partial}{\partial y}\log\frac{-6}{(y-x)^2} + \frac{-2}{y-x}$$

$$= -\frac{4}{y-x} - \frac{2}{y-x}$$

$$= -\frac{6}{y-x}.$$

so

$$\omega_i = \frac{1}{\left(-\frac{6}{(y-x_i)^2}\right)^{\frac{2}{3}}\left(\frac{-6}{(y-x_i)^2}\right)^{\frac{1}{3}}\left(\frac{-6}{(y-x)^2}\right)^{\frac{1}{3}}} \times \left(\frac{-6}{y-x} - \frac{-6}{y-x_i}\right)$$

$$= \frac{(y-x_i)^2(y-x)^{\frac{2}{3}}}{-6^{\frac{4}{3}}}\left(\frac{6x_i - 6x}{(y-x_i)(y-x)}\right)$$

$$= \frac{(y-x_i)(x-x_i)}{6^{\frac{1}{3}}(y-x)^{\frac{1}{3}}}$$

Each ω_i corresponding to a constant x_i is now a solution of the system of equations (26). However, we need to find three values of x_i that will form a fundamental system of (26) in ω_i. For simplicity we can drop the $6^{\frac{1}{3}}$ term in denominator, as a constant will cancel out in each equation of (26). Then let's try $x_1 = 1$, $x_2 = 0$ and $x_3 = -1$, yielding

$$\begin{cases} \omega_1 = \dfrac{(y-1)(x-1)}{(y-x)^{\frac{1}{3}}}, \\ \omega_2 = \dfrac{xy}{(y-x)^{\frac{1}{3}}}, \\ \omega_3 = \dfrac{(y+1)(x+1)}{(y-x)^{\frac{1}{3}}}. \end{cases}$$

3.4. THE CASE OF ONE RECTILINEAR AND TWO CURVED SCALES

In fact, they do constitute a fundamental system because (28) is met:

$$\Delta(\omega_1, \omega_2, \omega_3) = \begin{vmatrix} \frac{(y-1)(x-1)}{(y-x)^{\frac{1}{3}}} & -\frac{(y-1)(2x-3y+1)}{3(y-x)^{\frac{4}{3}}} & \frac{(x-1)(-3x+2y+1)}{3(y-x)^{\frac{4}{3}}} \\ \frac{xy}{(y-x)^{\frac{1}{3}}} & -\frac{y(-2x+3y)}{3(y-x)^{\frac{4}{3}}} & \frac{x(-3x+2y)}{3(y-x)^{\frac{4}{3}}} \\ \frac{(y+1)(x+1)}{(y-x)^{\frac{1}{3}}} & -\frac{(y+1)(-2x+3y+1)}{3(y-x)^{\frac{4}{3}}} & \frac{(x+1)(-3x+2y-1)}{3(y-x)^{\frac{4}{3}}} \end{vmatrix} = 2,$$

which is a non-zero constant. The algebraic details are omitted but straightforward. Each of the functions ω_1, ω_2 and ω_3 can also be confirmed as solutions to the system of equations in (26).

We can now follow the steps outlined earlier to find f_i and g_i of the standard determinant equation:

$$u = \frac{\omega_1}{\omega_3} = \frac{(y-1)(x-1)}{(y+1)(x+1)},$$

$$v = \frac{\omega_2}{\omega_3} = \frac{xy}{(y+1)(x+1)},$$

$$f_1(x) = -\frac{\frac{\partial v}{\partial y}}{\frac{\partial u}{\partial y}} = -\frac{\frac{x}{(y+1)^2(x+1)}}{\frac{2(x-1)}{(y+1)^2(x+1)}} = -\frac{x}{2(x-1)},$$

$$f_2(y) = -\frac{\frac{\partial v}{\partial x}}{\frac{\partial u}{\partial x}} = -\frac{\frac{y}{(y+1)(x+1)^2}}{\frac{2(y-1)}{(y+1)(x+1)^2}} = -\frac{y}{2(y-1)},$$

$$g_1(x) = uf_1(x) + v = \frac{(y-1)(x-1)}{(y+1)(x+1)}\left(-\frac{x}{2(x-1)}\right) + \frac{xy}{(y+1)(x+1)}$$
$$= \frac{x}{2(x+1)},$$

$$g_2(y) = uf_2(y) + v = \frac{(y-1)(x-1)}{(y+1)(x+1)}\left(-\frac{y}{2(y-1)}\right) + \frac{xy}{(y+1)(x+1)}$$
$$= \frac{y}{2(y+1)},$$

$$f_3(z) = -\frac{M\frac{\partial v}{\partial x} + \frac{\partial v}{\partial y}}{M\frac{\partial u}{\partial x} + \frac{\partial u}{\partial y}} = -\frac{-\frac{x^2}{y^2}\frac{y}{(x+1)^2(y+1)} + \frac{x}{(x+1)(y+1)^2}}{-\frac{x^2}{y^2}\frac{2(y-1)}{(x+1)^2(y+1)} + \frac{2(x-1)}{(x+1)(y+1)^2}}$$
$$= \frac{xy}{2(x+y)} = \frac{z}{2},$$

$$g_3(z) = uf_3(z) + v = \frac{(y-1)(x-1)}{(y+1)(x+1)}\left(\frac{xy}{2(x+y)}\right) + \frac{xy}{(y+1)(x+1)}$$
$$= \frac{xy}{2(x+y)} = \frac{z}{2}.$$

100   3. COMMENTARY

Therefore, the standard determinant form of the equation $z = \frac{xy}{x+y}$ is

$$\begin{vmatrix} -\frac{x}{2(x-1)} & \frac{x}{2(x+1)} & 1 \\ -\frac{y}{2(y-1)} & \frac{y}{2(y+1)} & 1 \\ \frac{z}{2} & \frac{z}{2} & 1 \end{vmatrix} = 0.$$

This nomogram is shown in Figure 3.9. Since $f_3(z) = g_3(z)$, the z scale runs diagonally, but the nomogram has been rotated 45° to maximize its size on the page. The x and y scales coincide and share the same numbering, although this procedure can also produce independent scales, of course. The sample isopleth shows the solution $z = 2.5$ for $x = 1.25$ and $y = -2.5$.

Once enlarged, this nomogram is fine for a narrow range of x and y values, and the linear z scale is excellent for precise readings over its entire range. This range can be shifted by multiplying x and y by a constant a, where the z scale is also multiplied by a. However, in addition to the narrow range restriction, there is very little opportunity to find a solution z when x and y are both positive or both negative.

We can cast this nomogram into new forms that may have more utility. One way to do this is to use a different fundamental system of w_i functions. Gronwall shows that if $\overline{w}_i$ forms a fundamental system, then

(31) $$\overline{w}_i = \alpha_i w_1 + \beta_i w_2 + \gamma_i w_3 \quad (i = 1, 2, 3).$$

This new fundamental system is associated with the same C, but a homographic transformation of the original determinant equation results.

Let's try to create a new fundamental system from linear combinations of the existing system:

$$\begin{cases} \overline{w}_1 = -\frac{1}{2}w_1 + (0)w_2 + \frac{1}{2}w_3 = \dfrac{x+y}{(y-x)^{\frac{1}{3}}} \\ \overline{w}_2 = (0)w_1 + (-1)w_2 + (0)w_3 = \dfrac{-xy}{(y-x)^{\frac{1}{3}}} \\ \overline{w}_3 = \frac{1}{2}w_1 - 2w_2 + \frac{1}{2}w_3 = \dfrac{1}{(y-x)^{\frac{1}{3}}} \end{cases}$$

3.4. THE CASE OF ONE RECTILINEAR AND TWO CURVED SCALES 101

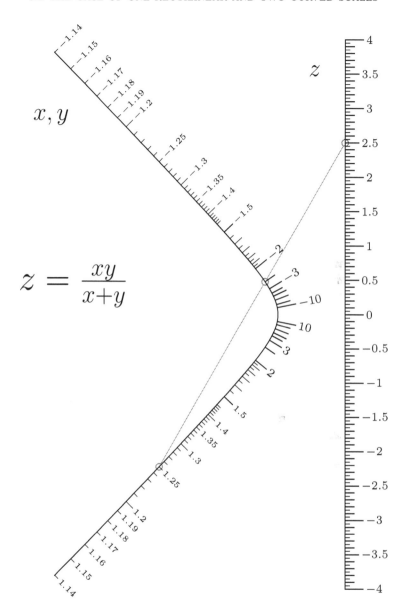

FIGURE 3.9. Nomogram for $z = \frac{xy}{x+y}$.

3. COMMENTARY

To verify that these $\bar{\omega}_i$ constitute a fundamental system, we can verify that (28) is met:

$$\Delta(\bar{\omega}_1, \bar{\omega}_2, \bar{\omega}_3) = \begin{vmatrix} \frac{(x+y)}{(y-x)^{\frac{4}{3}}} & \frac{4y-2x}{3(y-x)^{\frac{4}{3}}} & \frac{2y-4x}{3(y-x)^{\frac{4}{3}}} \\ -\frac{xy}{(y-x)^{\frac{1}{3}}} & \frac{y(2x-3y)}{3(y-x)^{\frac{4}{3}}} & \frac{x(3x-2y)}{3(y-x)^{\frac{4}{3}}} \\ \frac{1}{(y-x)^{\frac{1}{3}}} & \frac{1}{3(y-x)^{\frac{4}{3}}} & -\frac{1}{3(y-x)^{\frac{4}{3}}} \end{vmatrix} = 1,$$

which is a non-zero constant. Again, the algebraic details are omitted but straightforward. Each of the functions $\bar{\omega}_1$, $\bar{\omega}_2$ and $\bar{\omega}_3$ can also be confirmed as solutions to the system of equations in (26), although Gronwall has shown that this must be true if (28) is met and $\bar{\omega}_i$ are linear combinations of other ω_i that form a fundamental system of (26).

We again follow the steps to find f_i and g_i of this new standard determinant equation:

(3.4.4) $$u = \frac{\omega_1}{\omega_3} = x + y,$$

(3.4.5) $$v = \frac{\omega_2}{\omega_3} = -xy,$$

(3.4.6) $$f_1(x) = -\frac{\frac{\partial v}{\partial y}}{\frac{\partial u}{\partial y}} = -\frac{-x}{1} = x,$$

(3.4.7) $$f_2(y) = -\frac{\frac{\partial v}{\partial x}}{\frac{\partial u}{\partial x}} = -\frac{-y}{1} = y,$$

(3.4.8) $$g_1(x) = uf_1(x) + v = (x+y)x - xy = x^2,$$

(3.4.9) $$g_2(y) = uf_2(y) + v = (x+y)y - xy = y^2,$$

(3.4.10) $$f_3(z) = -\frac{M\frac{\partial v}{\partial x} + \frac{\partial v}{\partial y}}{M\frac{\partial u}{\partial x} + \frac{\partial u}{\partial y}} = -\frac{\frac{-x^2}{y^2}(-y)+(-x)}{\frac{-x^2}{y^2}(1)+1} = \frac{xy}{x+y} = z,$$

(3.4.11) $$g_3(z) = uf_3(z) + v = (x+y)\frac{xy}{x+y} - xy = 0.$$

This new standard determinant form of the equation $z = \frac{xy}{x+y}$ is

(3.4.12) $$\begin{vmatrix} x & x^2 & 1 \\ y & y^2 & 1 \\ z & 0 & 1 \end{vmatrix} = 0.$$

This nomogram is shown in Figure 3.10. The x and y scales again coincide and share the same numbering, but they lie along a parabola since $g_1(y) =$

$f_1^2(x)$ and $g_2(y) = f_2^2(x)$. The sample isopleth shows the solution $z = 4.67$ for $x = -3.5$ and $y = 2$.

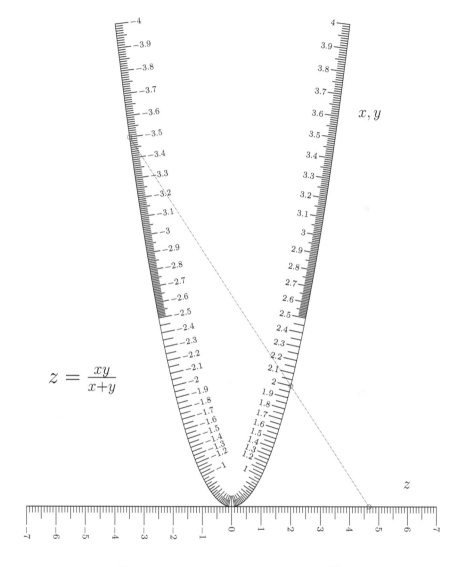

FIGURE 3.10. Nomogram for $z = \frac{xy}{x+y}$.

This nomogram has a wider range of x and y values, while the ranges of the x and y scales can be shifted as before. In the next section we will also see a way of shifting the x and y scales in different amounts and directions. This form of the nomogram also provides better opportunities to find a solution z when x and y are both positive or both negative.

This form of nomogram, where two scales are coincident along a single conic, is called a conical nomogram. They were discovered by J. Clark, who went on to create exquisite nomograms by studying their properties. For example, our original equation $z = \frac{xy}{x+y}$ can be written as $zx + zy - xy = 0$, but notice that when the determinant equation (3.4.12) is expanded it becomes $(x-y)(zx+zy-xy) = 0$. The extra factor $(x-y)$, or $(f_1 - f_2)$ generally, is one of the characteristics of this type of nomogram. Sections 4 and 6 of Gronwall's paper describe simplified procedures for equations that can be represented as conical nomograms, and we will return to the discussion of conical nomograms at those points. This also answers the lingering question of how it was possible to predict that $C = \frac{2}{y-x}$ was the common integral of Gronwall's system of equations (24) for $z = \frac{xy}{x+y}$.

3.5. The Case of $\frac{\partial^2 \log M}{\partial x \partial y} = 0$

As we have seen, Gronwall demonstrates in his Section 2 that

(50) $$\frac{\partial^2 \log M}{\partial x \partial y} = 0$$

comprises the necessary and sufficient condition so that equation $F(x, y, z) = 0$ can be reduced to the simple addition of functions of x, y, and z, as in

(51) $$\phi(x) + \psi(y) + \chi(z) = 0.$$

This seemingly simple equation corresponds to a wide variety of engineering and mathematical formulas. We have seen in Section 3.1 that these functions can be as complicated as needed, and with the use of logarithms (51) can be applied to multiplications and divisions of functions to various powers. It is therefore of much interest to investigate the families of nomograms possible for this equation, which Gronwall proceeds to do in his Section 4.

Gronwall describes a method at the beginning of this section to find the functions $\phi(x)$, $\psi(y)$, and $\chi(z)$ when an equation meets the test in (50). We have already looked into this procedure in Section 3.3, and so we will continue on with his classification of nomograms for this type of equation.

Gronwall remarks that a change of variables reduces any equation $\phi(x) + \psi(y) + \chi(z) = 0$ to

(3.5.1) $$x + y + z = 0,$$

and this certainly simplifies the presentation. With this form of the equation, Gronwall finds $M = -1$, $N = 0$, and $D = C$, reducing (23) to

(71) $$[\psi(x+2y) - \phi(2x+y)]C = \phi'(2x+y) + \psi'(x+2y)$$

He then separately considers four cases based on whether ϕ' and ψ' are zero:

I. $\phi' = \psi' = 0$;
II. $\phi' = 0 \quad \psi' \neq 0$;
III. $\phi' \neq 0 \quad \psi' = 0$;
IV. $\phi' \neq 0 \quad \psi' \neq 0$.

Each of these has sub-cases that arise in their study. The cases and sub-cases that Gronwall extracts are summarized in Table 3.1. For discussion purposes we have assigned a consistent sub-case numbering format for Case IV, something Gronwall did not do.

Here Gronwall manages to categorize all possible nomograms of the form $x + y + z = 0$, which we will now survey.

3.5.1. Case I. For $\phi' = \psi' = 0$ in (71), ϕ and ψ are constants $c_1 = c_2 = -\frac{1}{3}c$. This case is then subdivided into five cases (Iα1, Iα2, Iβ, Iγ1 and Iγ2) as shown in Table 3.1. Each gives rise to a particular family of nomograms as described below.

Case Iα1: For this case Gronwall derives the following determinant equation:

(75) $$\begin{vmatrix} 1 & x & 1 \\ 0 & y & 1 \\ \frac{1}{2} & -\frac{1}{2}z & 1 \end{vmatrix} = 0,$$

which yields the equation $x + y + z = 0$ on expansion.

The nomogram corresponding to this determinant appears in Figure 3.11. This is equivalent to Figure 3.6 derived in Section 3.3, but rotated 90°. This is the common parallel-scale nomogram for addition.

Case Iα2: Here Gronwall derives the following determinant equation:

(78) $$\begin{vmatrix} \frac{1}{x-x_0} & \frac{1}{(x-x_0)^2} & 1 \\ \frac{1}{y-y_0} & \frac{1}{(y-y_0)^2} & 1 \\ -\frac{1}{z-z_0} & 0 & 1 \end{vmatrix} = 0.$$

Case	ϕ'	ψ'	Sub-case	Sub-sub-case	Scale Layouts
I	0	0	Iα: $c = 0$	Iα1: $\chi = 0$	x, y, z rectilinear, concurrent
				Iα2: $\chi \neq 0$	x, y on conic, z rectilinear, tangent
			Iβ: $c > 0$		x, y on conic, z rectilinear, no intersection
			Iγ: $c < 0$	Iγ1: $\frac{1}{2}\chi^2 + 2c = 0$	x, y, z rectilinear, non-concurrent
				Iγ2: $\frac{1}{2}\chi^2 + 2c \neq 0$	x, y on conic, z rectilinear, intersects
II	0	$\neq 0$	IIα: $c = 0$	IIα1: $\chi = 0$	x, y, z rectilinear, concurrent
				IIα2: $\chi \neq 0$	y, z on conic, x rectilinear, tangent
			IIβ: $c > 0$		y, z on conic, x rectilinear, no intersection
			IIγ: $c < 0$	IIγ1: $\frac{1}{2}\chi^2 + 2c = 0$	x, y, z rectilinear, non-concurrent
				IIγ2: $\frac{1}{2}\chi^2 + 2c \neq 0$	y, z on conic, x rectilinear, intersects
III	$\neq 0$	0	IIIα: $c = 0$	IIIα1: $\chi = 0$	x, y, z rectilinear, concurrent
				IIIα2: $\chi \neq 0$	x, z on conic, y rectilinear, tangent
			IIIβ: $c > 0$		x, z on conic, y rectilinear, no intersection
			IIIγ: $c < 0$	IIIγ1: $\frac{1}{2}\chi^2 + 2c = 0$	x, y, z rectilinear, non-concurrent
				IIIγ2: $\frac{1}{2}\chi^2 + 2c \neq 0$	x, z on conic, y rectilinear, intersects
IV	$\neq 0$	$\neq 0$	IVα : *General*		Single Weierstrass's elliptic curve
			IVβ : *Degenerate*	IVβ1 : $g_2 > 0, g_3 > 0$	Single Weierstrass's elliptic curve
				IVβ2 : $g_2 > 0, g_3 < 0$	Single Weierstrass's elliptic curve
				IVβ3 : $g_2 = 0, g_3 = 0$	Single Weierstrass's elliptic curve

TABLE 3.1. Cases for $\frac{\partial^2 M}{\partial x \partial y} = 0$.

Here there is one arbitrary parameter $x_0 - y_0$ that can be set, where z_0 is found from the relationship

(3.5.2) $$x_0 + y_0 + z_0 = 0.$$

Again, Gronwall defines a *family* of nomograms as having the same value of C but allowing all homographic variations, while all equations obtained by varying *parameters* of a family constitutes a *class*.

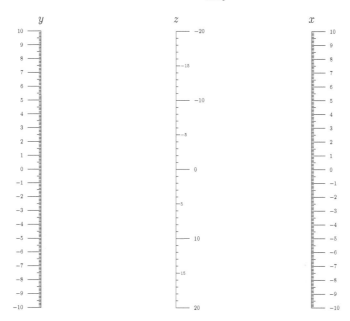

FIGURE 3.11.   Case Iα1 for $x + y + z = 0$.

Consider the conical nomogram in Figure 3.10 of Section 3.4 for the equation $z = \frac{xy}{x+y}$. This equation can be rearranged as $\frac{1}{x} + \frac{1}{y} - \frac{1}{z} = 0$. We can arrive at the simple form $x + y + z = 0$ if we make a variable substitution in which x and y are replaced with their reciprocals and z is replaced by the negative of its reciprocal. Making this substitution in (78) with $x_0 = y_0 = z_0$, we arrive immediately at the determinant equation (3.4.12) that we previously derived by means that were much more time-consuming. This also reveals another way of shifting the ranges of the x and y scales independently by introduction of x_0, y_0 and z_0 according to (3.5.2). Gronwall provides C for this case that applies as well to our determinant equation (3.4.12),

$$C = \frac{2}{y - x},$$

which is the assumed value of C in (3.4.2) that we tested against Gronwall's system of equations in Section 3.4.

Expanding the determinant in (78), we obtain the equation represented by the nomogram as

(3.5.3) $$\frac{(\overline{x - x_0} - \overline{y - y_0})(x + y + z)}{(x - x_0)^2(y - y_0)^2(z - z_0)} = 0,$$

where Gronwall's notation $\overline{x - x_0}$ and $\overline{y - y_0}$ simply denote $x - x_0$ and $y - y_0$. Notice again the extra factors that show up for nomograms with shared scales. This determinant equation cannot be obtained from that of (75) by a homographic transformation even for $x_0 = y_0 = z_0 = 0$, and as such it represents a new family of nomograms in which the x and y scales lie along a single conic, while the z scale is rectilinear and tangential to it.

The nomogram represented by the determinant in equation (78) for $x_0 = 5$, $y_0 = 1$ and $z_0 = -x_0 - y_0 = -6$ is shown in Figure 3.12. Now the x and y scales differ by $x_0 - y_0 = 4$, which allows independent selection of the ranges of these scales. However, we also need to have separate tick marks for them along the shared conic, which we can accommodate here by marking the x scale along the inside of the parabola and the y scale along the the outside. This dual marking is required in other conical nomograms when the two scales have different ranges, but for simplicity x_0 and y_0 are set equal in our later examples.

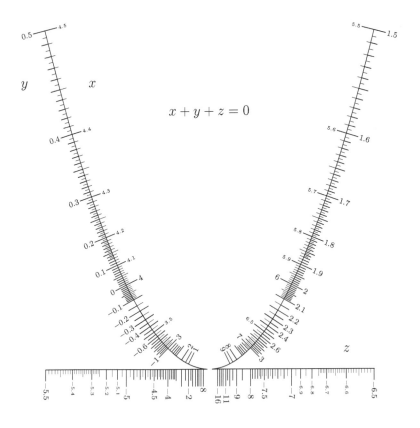

FIGURE 3.12. Case Iα2 for $x + y + z = 0$.

3.5. THE CASE OF $\frac{\partial^2 \log M}{\partial x \partial y} = 0$

Case Iβ: Gronwall also derives another form of a conical nomogram:

(80)
$$\begin{vmatrix} \cot a(x - x_0) & \cot^2 a(x - x_0) & 1 \\ \cot a(y - y_0) & \cot^2 a(y - y_0) & 1 \\ -\cot a(z - z_0) & -1 & 1 \end{vmatrix} = 0.$$

This determinant equation produces a family of conical nomograms in which the x and y scales lie along a single conic, while the z scale is rectilinear and does not intersect the conic. Here there are two parameters, $a > 0$ and $x_0 - y_0$. The nomogram represented by the determinant in equation (80) for $a = 0.05$ and $x_0 = y_0 = z_0 = 0$ is shown in Figure 3.13. Gronwall attributes the case $a = 1$ to Clark. Notice that the arguments of the cotangent terms must lie within the range $\pm\frac{\pi}{2}$ or the scales will wrap around to the other side of the parabola, limiting us here to values $x, y < 10\pi$.

Case Iγ1: Here Gronwall derives a family of nomogram containing a single parameter $a \neq 0$:

(82)
$$\begin{vmatrix} 0 & e^{ax} & 1 \\ 1 & e^{-ay} & 1 \\ \frac{a}{1-e^{az}} & 0 & 1 \end{vmatrix} = 0$$

A nomogram for $a = 0.1$ is shown in Figure 3.14. The cases for $a = \pm 1$ are attributed to d'Ocagne, but here Gronwall provides the general treatment for all values of a.

Case Iγ2: Yet another form of conical nomogram is derived in this case:

(84)
$$\begin{vmatrix} \coth a(x - x_0) & \coth^2 a(x - x_0) & 1 \\ \coth a(y - y_0) & \coth^2 a(y - y_0) & 1 \\ -\coth a(z - z_0) & -1 & 1 \end{vmatrix} = 0$$

This family of nomograms, which contains two parameters $a > 0$ and $x_0 - y_0$, consists of x and y scales lying along a single conic, while the z scale is rectilinear and intersects the conic in two points. Clark is credited with discovering this case for $a = 1$. A nomogram for $a = 0.1$ and $x_0 = y_0 = z_0 = 0$ is shown in Figure 3.15, which has also been scaled vertically by a factor of 5 to better fit the page. The ranges of the x and y scales are the same as they are in Figure 3.13. However the hyperbolic cotangent function, unlike the cotangent function, is not periodic, and so there are no limits on the scales other than being non-zero. As x and y scales increase, they approach the z scale, so the lower part of the overall parabolic curve lies outside the ranges of the scales

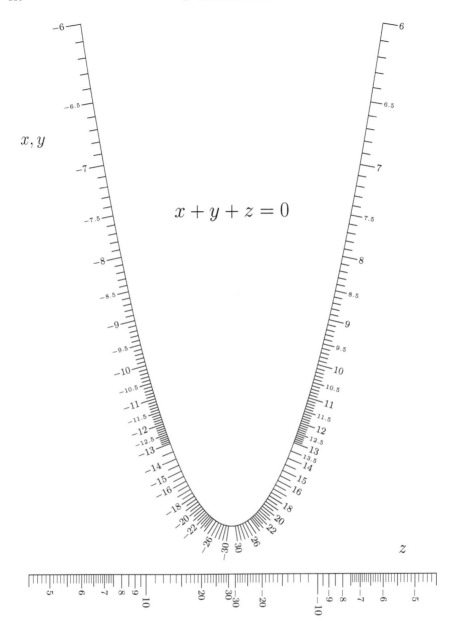

FIGURE 3.13. Case Iβ for $x+y+z=0$.

3.5. THE CASE OF $\frac{\partial^2 \log M}{\partial x \partial y} = 0$

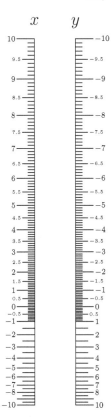

FIGURE 3.14. Case Iγ1 for $x + y + z = 0$.

along it. Nonetheless, this is what Gronwall means when he states that the z scale intersects the parabola in two points.

3.5.2. Case II. For $\phi' = 0$, $\psi' \neq 0$ in (71), ϕ is a constant c. This case is then subdivided into five cases (IIα1, IIα2, IIβ, IIγ1 and IIγ2) as shown in Table 3.1. Each gives rise to a particular type of nomogram. The values of C for IIα1 and IIα2 are the same as those for the earlier Iα1 and Iα2 cases. They are different for the other cases, but we have been focusing on the ultimate

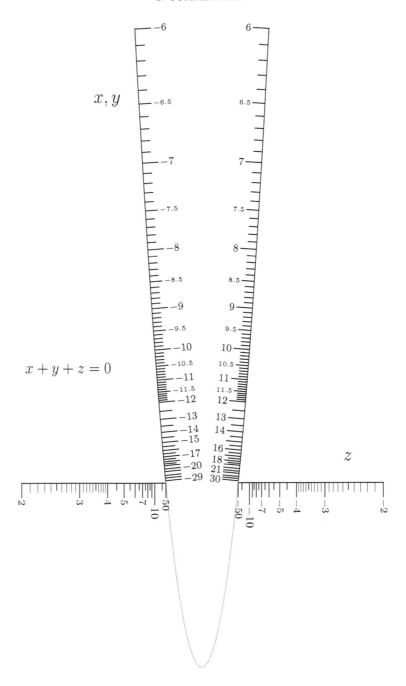

FIGURE 3.15. Case Iγ2 for $x + y + z = 0$.

determinate equations in this section rather than values of C. In all five sub-cases the families of nomograms match those of the corresponding cases for Case I if the x and z scales of the nomograms are swapped. So although this variable switch does not represent a homographic transformation for the IIβ, IIγ1 and IIγ2 sub-cases, and therefore these sub-cases represent additional new families of nomograms, we will not redraw the nomograms for this case.

3.5.3. Case III. For $\phi' \neq 0$, $\psi' = 0$ in (71), ψ is a constant c. This case is then subdivided into five cases once again (IIIα1, IIIα2, IIIβ, IIIγ1 and IIIγ2) as shown in Table 3.1. Each gives rise to a particular family of nomogram. Once again, the values of C for IIIα1 and IIIα2 are the same as those for the earlier Iα1 and Iα2 cases and are different for the other cases. In all five sub-cases the families of nomograms match those of the corresponding cases for Case I if the y and z scales of the nomograms are swapped. Again, this variable switch does not represent a homographic transformation for the IIIβ, IIIγ1 and IIIγ2 sub-cases, and therefore these sub-cases represent additional new families of nomograms, but we will not redraw the nomograms for the simple variable switch here.

3.5.4. Case IV. In Case IV Gronwall derives a novel form of nomogram, one that has received very little attention in the literature on nomography.[2] These nomograms consist of a single elliptic curve along which all three scales lie; an isopleth that cuts this curve in three places provides a solution to $x + y + z = 0$ or, of course, the more general equation $\phi(x) + \psi(y) + \chi(z) = 0$.

This elliptic curve is generated from Weierstrass's elliptic function, which is denoted by its own symbol, $\wp$. This function $\wp(u)$ is the inverse of the elliptic integral

$$(3.5.4) \qquad u = \int_\wp^\infty \frac{dx}{\sqrt{4x^3 - g_2 x - g_3}},$$

where g_2 and g_3 are constants.

The related elliptic curve is represented by the function given by Gronwall:

$$(3.5.5) \qquad \eta^2 = 4\xi^3 - g_2 \xi - g_3,$$

where ξ and η are used to differentiate the x and y Cartesian axes, respectively, from the x and y variables used in Equation (3.5.4).

Among other properties of an elliptic curve, the points along it form an additive abelian group, i.e., an operation can be defined in which the "sum" of two points on the curve is another point on the curve that lies on a line passing

---

[2]Epstein [**258**, ch.8] discusses this form in a chapter devoted to Gronwall.

through the first two points but located on the opposite side of the ξ-axis from them.[3] This additive property can be exploited to create a nomogram for the addition of three functions. While the additive property of elliptic curves was known by Weierstrass, Gronwall was the first to directly derive it as a standard form for nomograms of this type of equation.

An example for the equation $x + y + z = 0$ is shown in Figure 3.16 for the parameters $g_2 = 1200$ and $g_3 = -8500$. Other values of g_2 and g_3 produce figures that have greater separation at the pinched region, or even collapse that region as we will see later, but the selected values provide a convenient envelope for our example nomogram. The curve in this figure is rotated 90° counterclockwise, so the ξ-axis is vertical and the η-axis is horizontal. To eliminate the need for the isopleth to cross the ξ-axis, two sets of labels are used, one inside the curve and one outside the curve. An isopleth passing through two points anywhere on the curve then produces three label values that add to zero, two inside the curve and one outside or vice-versa. An isopleth tangent to the curve represents two identical values at that point.

The sample isopleth crossing the ξ-axis in Figure 3.16 connects the scale values of 0.2399, 0.5491 and -0.7890 if the outside value is taken in its middle, or 0.2399, -1.1881 and 0.9482 using the inside value in its middle. The isopleth above the ξ-axis connects the scale values of 0.2399, 0.6501 and -0.8900 if the outside value is taken in its middle, or 0.2399, -1.0871 and 0.8472 otherwise. Vertical isopleths cross two values that add to zero.

The scale value for any point on this type of nomogram is given by

$$(3.5.6) \qquad u = \pm \int_{\xi}^{\infty} \frac{d\xi}{\sqrt{4\xi^3 - g_2\xi - g_3}},$$

where u represents each of the functions $\phi(x)$, $\psi(y)$, and $\chi(z)$. In other words, for the simplest equation $x + y + z = 0$, the values of u on the curve apply equally to x, y and z.

The integral in (3.5.6) cannot be solved directly but must be numerically computed to the accuracy desired. Epstein discusses methods of approximating the integral in different ranges; today we have computer programs that can easily perform numerical integration.

In practice the positive scale values in the portion of the curve above the ξ-axis are first calculated from this integral. Now the difference in any two lower scale values is the negative of that for the upper scale, so the positive value scale that wraps below the ξ-axis is calculated at each ξ by adding the difference between the value at $\eta = 0$ and the value at ξ on the upper scale to

---

[3]See Epstein [**258**, pp. 121–124] for a short proof of this additive property (*Abel's proof*).

3.5. THE CASE OF $\frac{\partial^2 \log M}{\partial x \partial y} = 0$

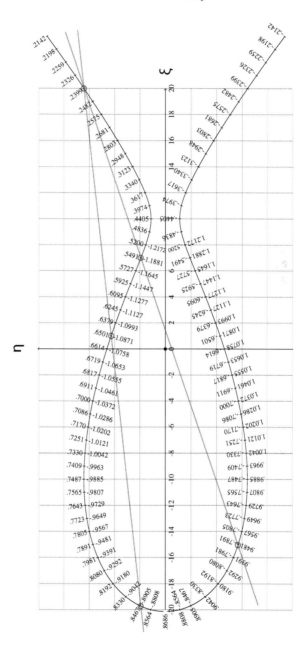

FIGURE 3.16. Weierstrass elliptic curve nomogram (rotated) for $x + y + z = 0$.

the value at $\eta = 0$. For example, in Figure 3.16 the value 0.9292 on the lower scale is found by adding $(0.8686 - 0.8080)$ to 0.8686, and in this way the entire positive value scale can be labeled. The negative scale values are simply the negatives of the values vertically opposite them on the curve.

Epstein also notes that a projective transformation can bring the points at infinity to zero and vice-versa. This transformation may be useful depending on the region of interest in the nomogram:

(3.5.7) $$\overline{f_i} = \frac{f_i}{g_i}, \quad \overline{g_i} = \frac{1}{g_i}$$

Assume we have the more general equation $\phi(x) + \psi(y) + \chi(z) = 0$. We can then assign scale values for $\phi(x)$, $\psi(y)$, and $\chi(z)$ that result in the original scale values x, y and z found above, and we can use the same curve. In other words, we label each point with the inverse of the function at that point. Notice in Figure 3.16 that when an isopleth intersects the upper curve at two or three points, the negative value is added to the two positive values. So $\phi(x)$, say, can be mapped to the negative values and $\psi(y)$ and $\chi(z)$ to the positive values. However, when an isopleth intersects the lower curve in two or three points, two negative values are added to a positive value, so we can continue to map $\phi(x)$ and $\psi(y)$ to the same negative and positive values as they wrap continuously around through the lower curve, but $\chi(z)$ must now be mapped to negative values. This means that the $\chi(z)$ scale values will be discontinuous when crossing the ξ-axis.

A rather fanciful nomogram for the oxygen consumption of rainbow trout, based again on the same curve, is shown in Figure 3.17.[4] Here the equation is $O_2 = KT^n W^m$, where

$K = 3.05 \times 10^{-4}$, $\quad n = 1.855$, $\quad m = -0.138$ $\quad$ for $T > 50$
$K = 1.90 \times 10^{-6}$, $\quad n = 3.130$, $\quad m = -0.138$ $\quad$ for $T < 50$

This equation is used in the design and calibration of aerators for rainbow trout racetracks at trout farms. Since there is a different equation for $T < 50$ than for $T > 50$, there are two sets of scales for T and W that map to the same O_2 scale, and this can be used to skirt the difficulty of the discontinuity of one of the positive value scales as it wraps around from the upper curve to the bottom curve. The equation for $T > 50$ is mapped on the upper curve, where T is mapped to a positive scale, W is mapped to a negative scale, and O_2 is mapped to a positive scale that wraps around the lower curve and continues as

---

[4]A version in which the nomogram scales have colors reminiscent of the actual scales of a rainbow trout, along with additional information on its construction, is also available [**275**].

3.5. THE CASE OF $\frac{\partial^2 \log M}{\partial x \partial y} = 0$ 117

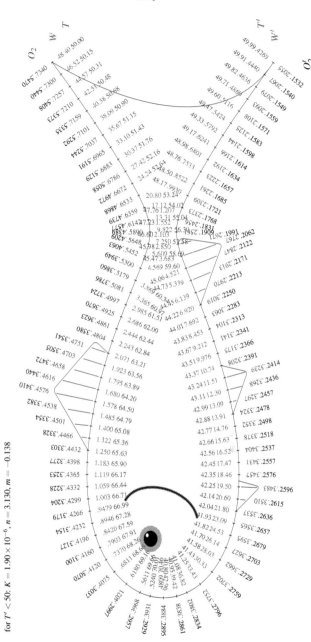

FIGURE 3.17. Zoomorphic Nomogram (rotated).

a positive scale. So for the case $T > 50$ the isopleth must cross two or three points above the ξ-axis since there are no W or T scales on the lower curve. This allows the third scale O_2 to wrap around and maintain positive values. The equation for $T < 50$ is implemented on the lower curve, with T' and W' mapped to negative value scales and O_2 mapped to a positive value scale that wraps around to the upper curve and continues as the positive value scale.

Now with the values $g_2 = 1200$ and $g_3 = -8500$ used in generating Figure 3.16, equation $\eta^2 = 4\xi^3 - g_2\xi - g_3$ of (3.5.5) has one real root (apparent from the single crossing point of the ξ-axis) and two complex roots. This is shown again in Figure 3.18.

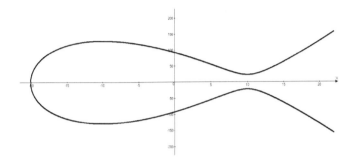

FIGURE 3.18. Weierstrass's elliptic curve with one real root.

For other values, say $g_2 = 1200$ and $g_3 = -4000$, Equation (3.5.5) has three real roots at $\xi = -18.8$, 3.47 and 15.3, producing the curve of Figure 3.19. Here an isopleth generally crosses both curves to connect the three values. If the isopleth tangentially touches the oval, it represents a repeated value. The isopleth could cross only the rightmost curve three times along the shallow curve if the scale were stretched enough to distinguish the crossing points.

Gronwall also discusses the degenerate cases of Weierstrass's elliptic curve in which two or all three of the roots of Equation (3.5.5) are equal. These cases offer nomograms in which all three scales lie along a single curve. To find the degenerate cases, we can set the first derivative of this equation to zero.

$$12\xi^2 - g_2 = 0,$$

$$\xi = \pm\sqrt{\frac{g_2}{12}}.$$

Note that the second equation implies $g_2 \geq 0$.

3.5. THE CASE OF $\frac{\partial^2 \log M}{\partial x \partial y} = 0$

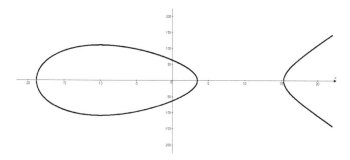

FIGURE 3.19. Weierstrass's elliptic curve with three real roots.

Substituting the negative value for ξ into Equation (3.5.5), we find
$$g_2^3 - 27g_3^2 = 0,$$
$$g_3 = \pm \left(\frac{g_2}{3}\right)^{\frac{3}{2}}.$$

Gronwall splits this into three cases: $g_3 > 0$ with $g_2 > 0$, $g_3 < 0$ with $g_2 > 0$, and $g_3 = 0$ with $g_2 = 0$. He then derives a new family of nomograms for each case. For our value of $g_2 = 1200$, we find with $g_3 = 8000$ that the oval curve collapses to a point where the open curve crosses the ξ-axis, denoting a triple real root of $\xi = 20$ as shown in Figure 3.20.

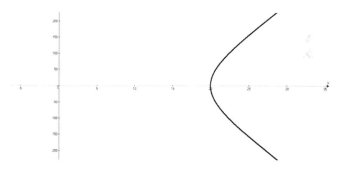

FIGURE 3.20. Weierstrass's elliptic curve with a triple real root.

Gronwall shows that this yields a determinant equation for $x + y + z = 0$ of

(107)
$$\begin{vmatrix} \frac{1}{\sin^2 a(x-x_0)} & \frac{\cos a(x-x_0)}{\sin^3 a(x-x_0)} & 1 \\ \frac{1}{\sin^2 a(y-y_0)} & \frac{\cos a(y-y_0)}{\sin^3 a(y-y_0)} & 1 \\ \frac{1}{\sin^2 a(z-z_0)} & \frac{\cos a(z-z_0)}{\sin^3 a(z-z_0)} & 1 \end{vmatrix} = 0,$$

and this produces the nomogram shown in Figure 3.21 for $a = 1$ and $x_0 = y_0 = z_0 = 0$. The sine and cosine functions are periodic, so we are limited to values less than $\pm\pi/2$. This is not at all the friendliest nomogram to use; the other ones we have seen are much more precise. In fact, the single-scale nomograms offered here differ only slightly in their appearance. But Gronwall is cataloging all the possibilities, and this interesting one can be a starting point for further transformations.

For our value of $g_2 = 1200$, we find with $g_3 = -8000$ that the two curves just touch, representing a double real root at that point. This plot is shown in Figure 3.22.

Gronwall shows that this yields a determinant equation for $x + y + z = 0$ of

(108)
$$\begin{vmatrix} \frac{1}{\sinh^2 a(x-x_0)} & \frac{\cosh a(x-x_0)}{\sinh^3 a(x-x_0)} & 1 \\ \frac{1}{\sinh^2 a(y-y_0)} & \frac{\cosh a(y-y_0)}{\sinh^3 a(y-y_0)} & 1 \\ \frac{1}{\sinh^2 a(z-z_0)} & \frac{\cosh a(z-z_0)}{\sinh^3 a(z-z_0)} & 1 \end{vmatrix} = 0,$$

and this produces the nomogram shown in Figure 3.23 for $a = 1$ and $x_0 = y_0 = z_0 = 0$. The shape of the nomogram is very similar to the previous one, but there are no longer any limits on the ranges.

For $g_2 = 0$ and $g_3 = 0$, the equation collapses to $\eta^2 = 4\xi^3$, a triple real root at $\xi = 0$. This corresponds to a determinant equation for $x + y + z = 0$ of

(109)
$$\begin{vmatrix} \frac{1}{(x-x_0)^2} & \frac{1}{(x-x_0)^3} & 1 \\ \frac{1}{(y-y_0)^2} & \frac{1}{(y-y_0)^3} & 1 \\ \frac{1}{(z-z_0)^2} & \frac{1}{(z-z_0)^3} & 1 \end{vmatrix} = 0,$$

and this produces the nomogram shown in Figure 3.24 for $x_0 = y_0 = z_0 = 0$. There is no longer a parameter a. The sample isopleth in this nomogram demonstrates that a tangent to the curve is a repeated value; here $1 + 1 - 2 = 0$.

We can make these latter nomograms more useable if we perform operations such as simple projective transformations on them. To illustrate, let's choose

3.5. THE CASE OF $\frac{\partial^2 \log M}{\partial x \partial y} = 0$

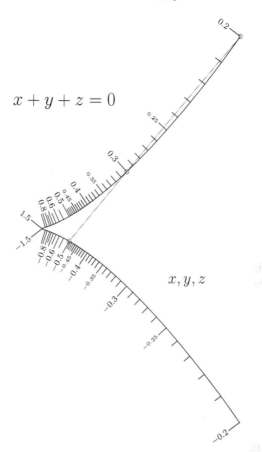

FIGURE 3.21. Single curve nomogram for a Weierstrass triple real root.

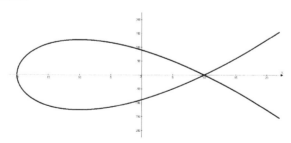

FIGURE 3.22. Weierstrass's elliptic curve with a double real root.

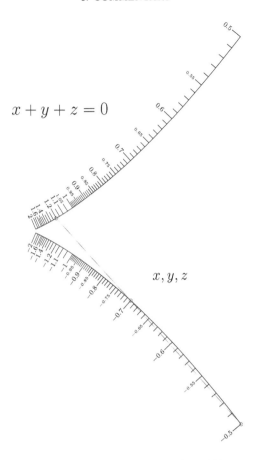

FIGURE 3.23. Single curve nomogram for a Weierstrass double real root.

a point of projection in Figure 3.24 just off the tip of the nomogram and just under the page, which should enhance the curvature in that region. From (109) we see that the tip at $x, y, z = \pm\infty$ lies at (0,0). If we choose a point of projection $(-0.01, 0, -0.1)$, the determinant equation is transformed according to (3.1.5) and we arrive at Figure 3.25. This is certainly a more graceful and more useful nomogram. By variable substitutions (e.g., replacing x by $3x$) and setting the values of x_0 and y_0 to values other than zero, a range of interest for each can be shifted into the region of greatest curvature. If $x_0 \neq y_0$, of course, we need individual scale labels for x, y and z.

Gronwall remarks that Clark has again anticipated him with some forms of these single-curve nomograms. We will see a few more of Clark's elegant

3.5. THE CASE OF $\frac{\partial^2 \log M}{\partial x \partial y} = 0$

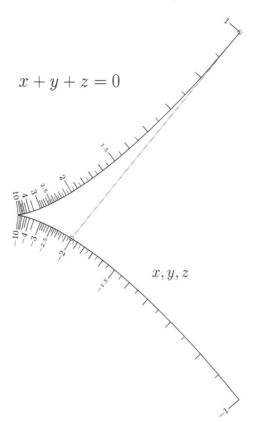

FIGURE 3.24. Single curve nomogram for a Weierstrass triple root at zero.

nomograms when we cover Gronwall's treatment of Clark's conical nomograms in the last section of his paper.

And so Gronwall has completed his remarkable catalogue of nomograms for the equation $x + y + z = 0$, or more generally for $\phi(x) + \psi(y) + \chi(z) = 0$. In concert with his test to determine if an equation can be put into such a form, he has provided an important asset to the adventurous nomographer.

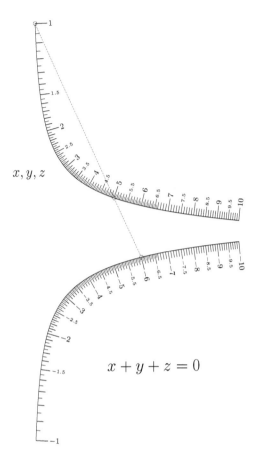

FIGURE 3.25. Nomogram of Figure 3.24 after simple projection.

3.6. The Case of One Curved and Two Rectilinear Scales

In Section 5, Gronwall continues his explorations of three-scale nomograms with an analysis of ones having one curved scale and two rectilinear scales as in Figure 3.26. Here he arbitrarily designates the scale of z as the curved one, but of course a change of variables will accommodate a situation where the x or y scale is curved. The curved scale can be located inside or outside the rectilinear scales.

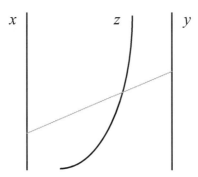

FIGURE 3.26. Nomogram with two rectilinear scales and one curved scale.

Gronwall derives an expression for C for this type of nomogram,

$$(110) \qquad C = -\frac{\frac{\partial}{\partial y}\frac{1}{M}\frac{\partial N}{\partial x}}{\frac{\partial^2 \log M}{\partial x \partial y}},$$

and the necessary and sufficient condition for this type of nomogram,

$$(111) \qquad \frac{\partial C}{\partial y} = \frac{\partial}{\partial x}(MC + N) = 0.$$

For equations that pass this condition, Gronwall provides a procedure for directly finding the elements in the standard determinant equation

$$(4) \qquad \begin{vmatrix} f_1(x) & g_1(x) & 1 \\ f_2(y) & g_2(y) & 1 \\ f_3(z) & g_3(z) & 1 \end{vmatrix} = 0.$$

As an example, consider the equation

$$(3.6.1) \qquad x - zy + z^2 = 0.$$

We can use implicit differentiation to find $\frac{\partial z}{\partial x}$ and $\frac{\partial z}{\partial y}$.

$$\frac{\partial}{\partial y}(x - zy + z^2) = \frac{\partial}{\partial y}(0)$$

$$0 - z - y\frac{\partial z}{\partial y} + 2z\frac{\partial z}{\partial y} = 0$$

$$\frac{\partial z}{\partial y} = \frac{z}{-y + 2z},$$

$$\frac{\partial}{\partial x}(x - zy + z^2) = \frac{\partial}{\partial x}(0)$$

$$1 - y\frac{\partial z}{\partial x} + 2z\frac{\partial z}{\partial x} = 0$$

$$\frac{\partial z}{\partial x} = -\frac{1}{-y + 2z}.$$

From (14),

(3.6.2)
$$\begin{cases} M = -\dfrac{\frac{\partial z}{\partial y}}{\frac{\partial z}{\partial x}} = z, \\ \\ N = \dfrac{\partial M}{\partial x} + \dfrac{1}{M}\dfrac{\partial M}{\partial y} \\ \\ = \dfrac{1}{-y+2z} + \dfrac{1}{z}\dfrac{-z}{-y+2z} \\ \\ = 0. \end{cases}$$

Now M is non-zero, so the nomogram cannot be represented by three rectilinear scales. We continue on to test for a curved z scale. Since $\frac{\partial N}{\partial x} = 0$, we get $C = 0$ from (110), and with (3.6.2) we have

$$\frac{\partial C}{\partial y} = \frac{\partial}{\partial x}(MC + N) = 0.$$

Since these conditions are met, we now know that the nomogram can be represented by rectilinear x and y scales and a curved z scale. To design it, we will follow Gronwall's procedure outlined in the last paragraph of his Section 5 to derive the elements f_i and g_i of the standard determinant equation (4).

First, we need to find $\varphi(x)$ and $\chi_1(z)$ from

(119a)
$$\left[\frac{\left(\frac{\partial z}{\partial x}\right)^2 \frac{\partial z}{\partial y}}{\frac{\partial^2 \log M}{\partial x \partial y}}\right]_y = \varphi(x)\chi_1(z),$$

3.6. THE CASE OF ONE CURVED AND TWO RECTILINEAR SCALES

where the subscript y denotes a solution in which y is eliminated by substitution, if necessary, from the original equation $x + zy + z^2 = 0$.

For the numerator,

$$\left(\frac{\partial z}{\partial x}\right)^2 \frac{\partial z}{\partial y} = \left(-\frac{1}{-y+2z}\right)^2 \frac{z}{-y+2z}$$

$$= \frac{z}{(-y+2z)^3}.$$

For the denominator,

$$\frac{\partial^2 \log M}{\partial x \partial y} = \frac{\partial}{\partial x}\left[\frac{\partial}{\partial y}\log(z)\right]$$

$$= \frac{\partial}{\partial x}\left(\frac{1}{z}\frac{\partial z}{\partial y}\right)$$

$$= \frac{\partial}{\partial x}\left(\frac{1}{z}\frac{z}{-y+2z}\right)$$

$$= \frac{\partial}{\partial x}\left(\frac{1}{-y+2z}\right)$$

$$= -\frac{2\frac{\partial z}{\partial x}}{(-y+2)^2}$$

$$= \frac{2}{(-y+2z)^3}.$$

Therefore,

$$\varphi(x)\chi_1(z) = \frac{\frac{z}{(-y+2z)^3}}{\frac{2}{(-y+2z)^3}} = \frac{z}{2},$$

so we have

$$\varphi(x) = 1,$$
$$\chi_1(z) = \frac{z}{2}.$$

Second, we find $\psi(x)$ and $\chi_2(z)$ from

(119b)
$$\left[\frac{\frac{\partial z}{\partial x}\left(\frac{\partial z}{\partial y}\right)^2}{\frac{\partial^2 \log M}{\partial x \partial y}}\right]_x = \psi(y)\chi_2(z).$$

We find in a similar manner as above,
$$\psi(y)\chi_2(z) = -\frac{z^2}{2},$$
so we have
$$\psi(y) = 1,$$
$$\chi_1(z) = -\frac{z^2}{2}.$$

Now we find any solution whatsoever for f_1 in

(121) $$-\frac{\varphi}{\frac{\varphi'}{\varphi} + \left(\frac{\partial \log M}{\partial x}\right)_y} = \frac{1}{f_1} + \frac{g_3'}{f_3' g_3 - g_3' f_3}.$$

Gronwall derives this relationship and and shows that the right side of this equation can always be expressed as the sum of this function of x and function of z. Using the substitution $\frac{z^2+x}{z}$ for y found by rearranging the original equation (3.6.1), we have

$$-\frac{\varphi}{\frac{\varphi'}{\varphi} + \left(\frac{\partial \log M}{\partial x}\right)_y} = -\frac{1}{0 + \left(\frac{-1}{z(-y+2z)}\right)_y}$$
$$= (z(-y+2z))_y$$
$$= z\left(-\frac{z^2+x}{z} + 2z\right)$$
$$= -x + z^2,$$

and we find by comparison to (121),

$$\boxed{f_1 = -\frac{1}{x}}$$

Similarly, we find any solution g_2 in

(122) $$-\frac{\psi}{\left(\frac{\partial \log M}{\partial y}\right)_x - \frac{\psi'}{\psi}} = \frac{1}{g_2} + \frac{g_3'}{f_3' g_3 - g_3' f_3},$$

where Gronwall again assures us of the existence of this expression for g_2. Here

$$-\frac{\psi}{\left(\frac{\partial \log M}{\partial y}\right)_x - \frac{\psi'}{\psi}} = -\frac{1}{\left(\frac{1}{-y+2z} - \frac{0}{1}\right)_x}$$
$$= -(-y+2z)_x$$
$$= y - 2z,$$

3.6. THE CASE OF ONE CURVED AND TWO RECTILINEAR SCALES

and we find by comparison to (122),

$$\boxed{g_2 = \frac{1}{y}}$$

The next step is to set

$$\boxed{\begin{array}{c} f_2 = 0 \\ g_1 = 0 \end{array}}$$

At this point we have the determinant equation

$$\begin{vmatrix} -\frac{1}{x} & 0 & 1 \\ 0 & \frac{1}{y} & 1 \\ f_3 & g_3 & 1 \end{vmatrix} = 0.$$

Then we calculate f_3 and g_3 from

(15) $$f_3(z) = -\frac{M\frac{\partial v}{\partial x} + \frac{\partial v}{\partial y}}{M\frac{\partial u}{\partial x} + \frac{\partial u}{\partial y}}$$

and

(7) $$g_3(z) = u f_3(z) + v,$$

where we can use either of these relationships to determine u and v:

(8) $$\begin{cases} u = \dfrac{g_1(x) - g_2(y)}{f_1(x) - f_2(y)}, \\ v = \dfrac{f_1(x)g_2(y) - f_2(y)g_1(x)}{f_1(x) - f_2(y)}, \end{cases}$$

or

(9) $$\begin{cases} g_1(x) = u f_1(x) + v, \\ g_2(y) = u f_2(y) + v. \end{cases}$$

Using the system of equations (9), we have

$$\begin{cases} 0 = u\left(-\dfrac{1}{x}\right) + v, \\ \dfrac{1}{y} = u(0) + v, \end{cases}$$

and we quickly arrive at $v = \frac{1}{y}$ and $u = \frac{x}{y}$. Then from (15), we have

$$f_3(z) = -\frac{z(0) + (-\frac{1}{y^2})}{z\frac{1}{y} + (-\frac{x}{y^2})}$$

$$= -\frac{1}{-zy + x},$$

but since $-zy + x = -z^2$ from the original equation (3.6.1),

$$\boxed{f_3(z) = \frac{1}{z^2}}$$

Also, from (7) we have

$$g_3(z) = \frac{x}{y}\left(\frac{1}{z^2}\right) + \frac{1}{y}$$

$$= \frac{x + z^2}{z^2 y},$$

but since $x + z^2 = zy$ from the original equation (3.6.1),

$$\boxed{g_3(z) = \frac{1}{z}}$$

We now have the complete determinant equation in standard form,

(3.6.3)
$$\begin{vmatrix} -\frac{1}{x} & 0 & 1 \\ 0 & \frac{1}{y} & 1 \\ \frac{1}{z^2} & \frac{1}{z} & 1 \end{vmatrix} = 0.$$

The determinant can be expanded to confirm that this equation is equivalent to the original equation $x - zy + z^2 = 0$.

Now that the determinant is in standard form, we can construct the nomogram. Here the x scale is horizontal with a scaling of $-\frac{1}{x}$ and the the y scale is vertical with a scaling of $\frac{1}{y}$. The z scale is a parabola opening to the right, as $f_3 = g_3^2$ maps to the general curve $x = y^2$. To achieve a larger page layout, however, we swap the first two columns of the determinant so the parabola opens upward, yielding

(3.6.4)
$$\begin{vmatrix} 0 & -\frac{1}{x} & 1 \\ \frac{1}{y} & 0 & 1 \\ \frac{1}{z} & \frac{1}{z^2} & 1 \end{vmatrix} = 0.$$

The nomogram for this determinant equation is shown in Figure 3.27. The sample isopleth provides the two real roots −0.372 and 0.672 for $x = -0.25$ and $y = 0.3$.

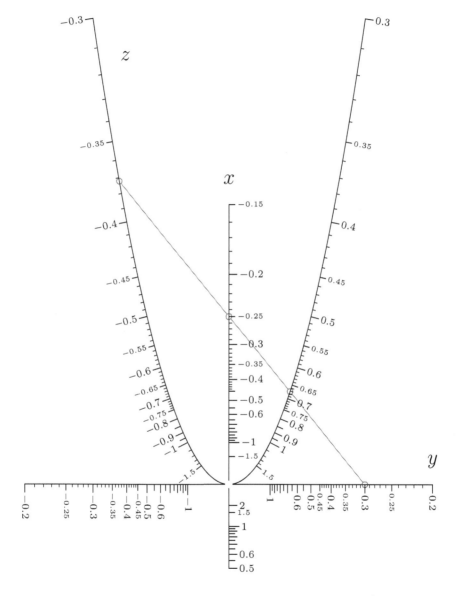

FIGURE 3.27. Nomogram for the equation $x - zy + z^2 = 0$.

Now this might seem to be a great deal of work to arrive at a standard determinant equation for such a simple original equation. However, Gronwall has provided a prescriptive means to test whether this type of standard determinant equation exists for our original equation, and then to ultimately find the final elements of the determinant. There exist listings of standard determinant equations for some forms of equation, but Gronwall provides a necessary and sufficient condition that encompasses all equations of this type.

Here we have directly found the elements f_i and g_i of the standard determinant, which we knew to exist after the equation passed the tests for a nomogram with one curved and two rectilinear scales. The general method presented in Section 3.1 produces a determinant equation, but often a great deal of educated guesswork and trial-and-error is required to convert this determinant equation into standard determinant form, if it is even possible to do so. Gronwall presents a significant advancement that becomes more apparent when the original equation is more complicated. Again, modern symbolic algebra programs can assist in performing these operations for more complicated equations.

Our example equation is more general than it might appear. If there is a coefficient other than unity in front of x^2 in an equation of this type, the entire equation can be divided by that number to arrive at an equation in our form. The ranges of the scales can also be changed by multiplying the equation $x - zy + z^2 = 0$ by a constant; for example, an equivalent equation is $(9x) - (3y)(3z) + (3z)^2 = 0$. Beyond that, our simple equation form $x - zy + z^2 = 0$ encompasses a vast range of equations because the variables x, y and z can be replaced by functions $f(x)$, $f(y)$ and $f(z)$. There are many equations in the sciences that exhibit this relationship.

Interestingly, an isopleth through two roots on the parabolic z scale will intersect the x scale at the product of two numbers, and the y scale as the sum. This happens because two solutions a and b imply $(z-a)(z-b) = 0$. Expanding this and equating terms to the original equation, we find that $x = ab$ and $y = a+b$. The sample isopleth demonstrates this, as $-0.372 \times 0.672 = -0.25$ and $-0.372 + 0.672 = 0.3$. In regard to the summation property, this nomogram bears a relationship to the conical nomograms in Section 3.5 for the the sum $x + y + z = 0$, or $-z = x + y$. For example, our determinant equation (3.6.4) can be converted to the determinant equation of Case $I\alpha 2$ in (78), drawn as a nomogram in Figure 3.12, if the y row were assigned to the variable $-z$ and the z row were split into x and y scales.

Once we have the determinant in standard form, of course, we can perform general transformations to produce homographic equivalents. An open conic, such as the parabola in our nomogram or a hyperbola in other nomograms,

3.6. THE CASE OF ONE CURVED AND TWO RECTILINEAR SCALES

can be converted into a closed conic (an ellipse or circle) through a projective transformation. In the era before computers, circular nomograms offered significant benefits for the draftsman; beyond that, circular nomograms constrain the entire range of the variable along a scale of finite rather than infinite length. They are also strikingly elegant creations.

We saw earlier that a projective transformation is effected by replacing each f_i and g_i of the determinant with new forms dependent on the chosen point of projection $P(x_P, y_P, z_P)$ as proscribed in (3.1.5). Assuming a projection point at $(-1, 0, 1)$, the standard determinant equation (3.6.3) transforms into

$$\begin{vmatrix} -\frac{1}{x-1} & 0 & 1 \\ 0 & \frac{1}{y} & 1 \\ \frac{1}{z^2+1} & \frac{z}{z^2+1} & 1 \end{vmatrix} = 0.$$

Swapping the first two columns provides a circle sitting upright on the x-axis:

$$\begin{vmatrix} 0 & -\frac{1}{x-1} & 1 \\ \frac{1}{y} & 0 & 1 \\ \frac{z}{z^2+1} & \frac{1}{z^2+1} & 1 \end{vmatrix} = 0.$$

Here we have parametric equations for the scale of z,

$$\begin{cases} f_3 = \dfrac{z}{z^2+1}, \\ g_3 = \dfrac{1}{z^2+1}, \end{cases}$$

and eliminating z we have the relation $f_3^2 + (g_3 - 1)^2 = 1$, a circle of radius 1 centered at $(0, 1)$. This form of the nomogram is shown in Figure 3.28.

This nomogram provides a very good visual model of the behavior of the equation $x - yz + z^2 = 0$. An isopleth through a value of x and a value of y will just touch the z circle if the discriminant $y^2 - 4x$ from the quadratic formula is 0, where the point on circle is the repeated real root. The sample isopleth shows the repeated real root 0.5 for $x = 0.25$ and $y = 1.0$. An isopleth will miss the circle if no real roots exist ($y^2 - 4x < 0$). Otherwise, the isopleth crosses the circle at the values of the two real roots of the equation.

As in the parabolic form of the nomogram, an isopleth through two values on the circle will intersect the x scale at the product of two numbers, and on the y scale as the sum. The sample isopleth demonstrates this, as $0.5^2 = 0.25$ and $0.5 + 0.5 = 1.0$.

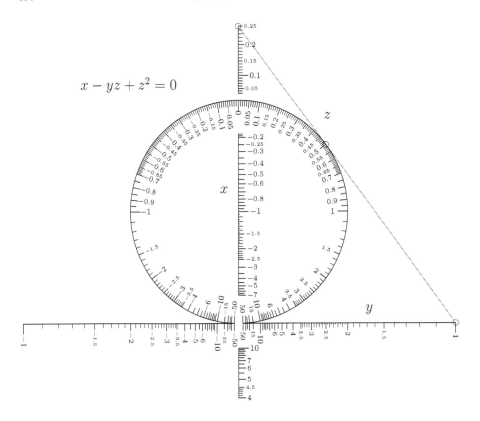

FIGURE 3.28. Circular nomogram for the equation $x - zy + z^2 = 0$.

It is possible to find complex roots of the equation as well with this nomogram. It turns out that if the variable x in the original equation $x - yz + z^2 = 0$ is replaced by $\frac{y^2}{2} - x$, the quadratic formula transforms from

$$z = \frac{y}{2} \pm \frac{\sqrt{y^2 - 4x}}{2} \quad \text{to} \quad z = \frac{y}{2} \pm \frac{\sqrt{4x - y^2}}{2}.$$

If the roots are complex, or in other words if $y^2 - 4x$ is negative, the second form negates this term. So we replace x with $\frac{y^2}{2} - x$ and find the corresponding real roots from the nomogram. Then the complex roots are complex conjugates $a \pm bi$, where we see from the quadratic formulas above that $a = \frac{y}{2}$ and $b = z_0 - a$ for either real root z_0 found from the nomogram.

Say we have the equation $0.295 - 0.3z + z^2 = 0$. An isopleth through the values $x = 0.295$ and $y = 0.3$ in Figure 3.27 falls beneath the parabola of

z values, so the equation has no real roots. We replace x by $\frac{0.3^2}{2} - 0.295 = -0.25$, and we find as before that our sample isopleth through these values provides the real roots 0.672 and -0.372. Therefore, $a = \frac{0.3}{2} = 0.15$ and $\pm b = \pm(0.672 - 0.15) = \pm(-0.372 - 0.15) = \pm 0.522$, so the complex roots are $0.15 \pm 0.522i$. The equation $5 - 2z + z^2 = 0$ falls below the circle in Figure 3.28, but an isopleth through $y = 2$ and $x = \frac{2^2}{2} - 5 = -3$ provides real roots of 3 and -1. Therefore, the complex roots are $1 \pm 2i$.

3.7. Clark's Nomograms

In the final Section 6 of his paper, Gronwall focuses on the conical nomograms of Clark that have so often come up in his work [187][188][191][192][269, pp. 122–157][230, pp. 228–244][259, pp. 155–161, 242–264][255][256][234][260, pp. 1–21]. As we have seen, in such a nomogram two of the scales (assumed to be the x and y scales) lie along a single conic, while the third scale (the z scale) is rectilinear or curved. In the general case, a curved z scale does not lie on the conic.

By variable substitution we can simplify any conical nomogram into the standard determinant equation,

(123) $$\begin{vmatrix} f_1(x) & f_1^2(x) & 1 \\ f_2(y) & f_2^2(y) & 1 \\ f_3(z) & g_3(z) & 1 \end{vmatrix} = 0.$$

Gronwall provides a test to identify whether an equation can be drawn as a conical nomogram. If we assume a value of C from

(130) $$C = -\frac{\frac{\partial}{\partial y} \frac{1}{M} \frac{\partial N}{\partial x}}{\frac{\partial^2 \log M}{\partial x \partial y}},$$

then this C must satisfy

(131) $$\begin{cases} \dfrac{\partial C}{\partial y} = \dfrac{\partial}{\partial x}(MC + N) = \dfrac{\partial D}{\partial x} = 0, \\ \dfrac{\partial^2 C}{\partial x \partial y} = C \dfrac{\partial C}{\partial y}. \end{cases}$$

Now in section 3.4 we considered the equation

(3.4.1) $$z = \frac{xy}{x+y}.$$

We find from (14),

$$\begin{cases} M = -\dfrac{\frac{\partial z}{\partial y}}{\frac{\partial z}{\partial x}} = -\dfrac{\frac{x^2}{(x+y)^2}}{\frac{y^2}{(x+y)^2}} \\ = -\dfrac{x^2}{y^2}, \\ \dfrac{\partial^2 \log M}{\partial x \partial y} = 0. \end{cases}$$

This last equation prevents us from using the test (131) for a conical nomogram, as there would be a zero in the denominator. Gronwall states that in this case the z scale is rectilinear, which we indeed found in Figure 3.10. This situation occurs for many equations, so we will will turn to Clark, who laid out forms of equations for conical nomograms, to verify that our equation meets the criteria.

Clark found that equations in the form

(3.7.1) $$f_1 f_2 A_3 + (f_1 + f_2) B_3 + C_3 = 0$$

can be represented by the determinant

(3.7.2) $$\begin{vmatrix} f_1 & f_1^2 & 1 \\ f_2 & f_2^2 & 1 \\ -B_3 & C_3 & A_3 \end{vmatrix} = 0,$$

where A_3, B_3 and C_3 are functions of z. Of course, we can divide the elements in the third row by A_3 to get the determinant equation in standard form. If you actually expand the determinant you will get the equation $(f_2 - f_1)(f_1 f_2 A_3 + (f_1 + f_2)B_3 + C_3) = 0$. This satisfies Clark's equation but consists of an extra term $(f_2 - f_1)$, corresponding to the $(x-y)$ term we saw earlier for our example. And intuitively this makes some sense, since $x - y = 0$ at every point when the two scales are identical.

Expanding our equation (3.4.1) we have $xy - (x+y)z = 0$, and we find it matches Clark's form $f_1 f_2 A_3 + (f_1 + f_2) B_3 + C_3 = 0$ for $f_1 = x$, $f_2 = y$, $A_3 = 1$, $B_3 = z$ and $C_3 = 0$. So we know it can be represented by a conical nomogram with the determinant equation we had discovered earlier in (3.4.12),

$$\begin{vmatrix} x & x^2 & 1 \\ y & y^2 & 1 \\ z & 0 & 1 \end{vmatrix} = 0.$$

3.7. CLARK'S NOMOGRAMS

Now we can use Gronwall's formula for C for this family of conical nomograms,

(124) $$C = \frac{f_1''}{f_1'} + \frac{2f_1'}{f_2 - f_1},$$

(3.7.3) $$= \frac{2}{y - x},$$

which in Section 3.4 was the expression that we tested against Gronwall's system of equations in C and found to hold for our equation $z = \frac{xy}{x+y}$. In that section we proceeded through a series of steps to find the elements of the standard determinant equation, while here we directly found them from the equation form. The intermediate values of u and v we calculated from C in (3.4.4) also match the expressions $u = f_1 + f_2 = x + y$ and $v = -f_1 f_2 = -xy$ shown in (124) of this section of Gronwall's paper. Again, this conical nomogram is shown in Figure 3.10.

Clark also considered another equation that matches the form (3.7.1),

(3.7.4) $$f_3 = \frac{f_1 + f_2}{1 - f_1 f_2}$$

or

(3.7.5) $$f_1 f_2 f_3 + f_1 + f_2 - f_3 = 0,$$

Equation (3.7.4) has the form of the tangent addition formula,

(3.7.6) $$\tan(a + b) = \frac{\tan a + \tan b}{1 - \tan a \tan b}.$$

From (3.7.5) we find $A_3 = f_3$, $B_3 = 1$ and $C_3 = -f_3$. Placing these into the standard determinant (3.7.2) and dividing the third row by f_3, we have

$$\begin{vmatrix} f_1 & f_1^2 & 1 \\ f_2 & f_2^2 & 1 \\ -\frac{1}{f_3} & -1 & 1 \end{vmatrix} = 0.$$

After substituting the tangent functions from (3.7.6), we have the conical nomogram drawn in Figure 3.29.

We have seen in our previous Section 3.6 that a conic can be transformed into a different conic by a projective transformation. Clark provides an alternative for his conical determinant equation (3.7.2) that converts the parabola

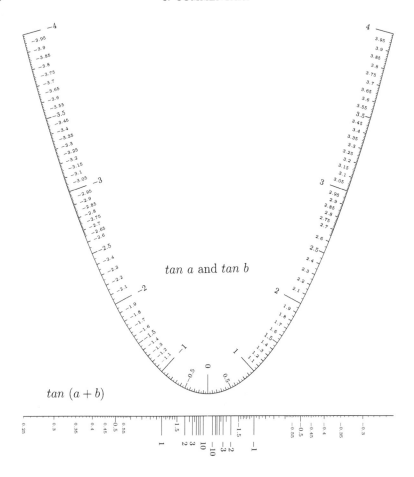

FIGURE 3.29. Conical nomogram for $\tan(a+b) = \frac{\tan a + \tan b}{1 - \tan a \tan b}$.

into a circle:

$$\begin{vmatrix} \frac{(f_1^2+1)\cos\alpha - 2f_1}{f_1^2+1-2f_1\cos\alpha} & \frac{(f_1^2-1)\sin\alpha}{f_1^2+1-2f_1\cos\alpha} & 1 \\ \frac{(f_2^2+1)\cos\alpha - 2f_2}{f_2^2+1-2f_2\cos\alpha} & \frac{(f_2^2-1)\sin\alpha}{f_2^2+1-2f_2\cos\alpha} & 1 \\ \frac{(A_3+C_3)\cos\alpha + 2B_3}{A_3+C_3+2B_3\cos\alpha} & \frac{(C_3-A_3)\sin\alpha}{A_3+C_3+2B_3\cos\alpha} & 1 \end{vmatrix} = 0.$$

The parameter α determines the angle on the circle of the corresponding opening of the parabola. For $\alpha = 45°$ we have the nomogram shown in Figure 3.30. A value $\alpha = 90°$ would produce zeros in the denominators, but $\alpha = 89.99°$ produces the nomogram in Figure 3.31.

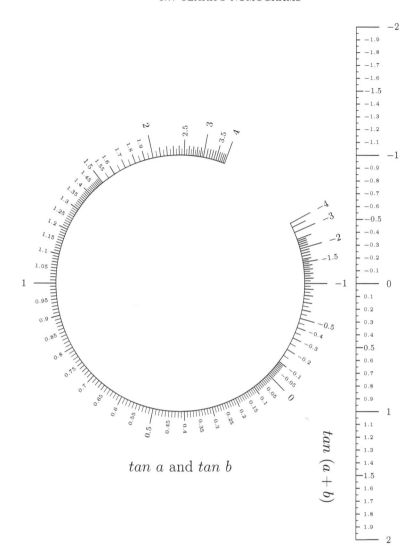

FIGURE 3.30. Circular nomogram for $\tan(a + b) = \frac{\tan a + \tan b}{1 - \tan a \tan b}$, $\alpha = 45°$.

Gronwall attributes to Clark some of the single-curve nomograms derived in Section 4 of his paper. We now look at these other single-scale nomograms derived by Clark for comparison.

In a flash of insight, it occurred to Clark that if the extra factor $(f_1 - f_2)$ is associated with nomograms in which the f_1 and f_2 scales are aligned, then a different factor might cause all three scales to align. He found this factor to

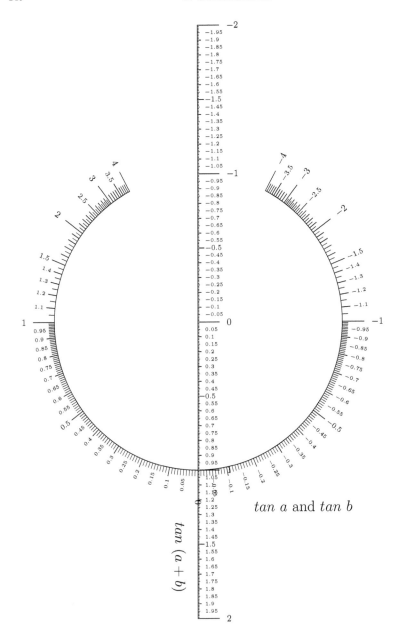

FIGURE 3.31. Circular nomogram for $\tan(a + b) = \frac{\tan a + \tan b}{1 - \tan a \tan b}$, $\alpha = 89.99°$.

be $(f_1 - f_2)(f_2 - f_3)(f_3 - f_1)$ when applied to general equation form

(3.7.7) $\qquad f_1 f_2 f_3 + A(f_1 f_2 + f_2 f_3 + f_1 f_3) + B(f_1 + f_2 + f_3) + C = 0,$

and this produces the determinant equation

(3.7.8)
$$\begin{vmatrix} \frac{f_1+A}{f_1^3+C} & \frac{f_1^2-B}{f_1^3+C} & 1 \\ \frac{f_2+A}{f_2^3+C} & \frac{f_2^2-B}{f_2^3+C} & 1 \\ \frac{f_3+A}{f_3^3+C} & \frac{f_3^2-B}{f_3^3+C} & 1 \end{vmatrix} = 0.$$

Equation form (3.7.7) can be separated into three classes of nomograms whose basic forms are

(3.7.9)
$$\begin{cases} \dfrac{1}{f_1} + \dfrac{1}{f_2} + \dfrac{1}{f_3} = 0 & \text{cuspidal form} \\ f_1 f_2 f_3 - (f_1 + f_2 + f_3) = 1 & \text{acnodal form} \\ f_1 f_2 f_3 = 1 & \text{crunodal form} \end{cases}$$

The cuspidal form is the harmonic relation we have seen in Gronwall's paper. Gronwall derives this family of nomograms from Weierstrass's elliptic curve for the case of a triple root at $\xi = 0$. This nomogram was shown in Figure 3.24. Here Gronwall is considering the equation $x + y + z = 0$, but the elements of his determinant equation in (109) apply to the harmonic relation when the variables are replaced by their reciprocals.

The equation $\frac{1}{f_1} + \frac{1}{f_2} + \frac{1}{f_3} = 0$ can be written as $(f_1 f_2 f_3)^{-1}(f_1 f_2 + f_2 f_3 + f_1 f_3)$, which corresponds to $A = 1$, $B = 0$ and $C = -f_1 f_2 f_3$ in Clark's general form (3.7.7). Inserting these values into the general determinant equation (3.7.8) and performing operations to return it to standard form yields Clark's determinant equation for the cuspidal form defined in the first equation of (3.7.9):

(3.7.10)
$$\begin{vmatrix} f_1^2 & f_1^3 & 1 \\ f_2^2 & f_2^3 & 1 \\ f_3^2 & f_3^3 & 1 \end{vmatrix} = 0.$$

Here the elements f_1, f_2 and f_3 are the reciprocals of those in Gronwall's equation (109), but the form is identical. This nomogram for $f_1 = x$, $f_2 = y$ and $f_3 = -z$, is drawn in Figure (3.32). The values of x and y are always those crossed on the same half of the curve due to the asymmetry of the equation; if f_3 were not negated, half of the curve would have negative values.

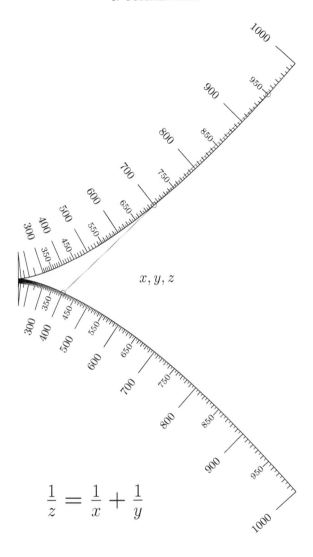

FIGURE 3.32. Cuspidal nomogram for the harmonic relation.

Now we can write the tangent addition formula (3.7.6) in the form $f_1 f_2 f_3 - (f_1 + f_2 + f_3) = 0$ for $f_1 = \tan a$, $f_2 = \tan b$ and $f_3 = -\tan(a+b)$. This corresponds to the second equation in (3.7.9), and to $A = 0$, $B = -1$ and

3.7. CLARK'S NOMOGRAMS

$C = 0$ in the general determinant equation (3.7.8), and we end up with

$$\begin{vmatrix} f_1 & \frac{f_1}{f_1^2+1} & 1 \\ f_2 & \frac{f_2}{f_2^2+1} & 1 \\ f_3 & \frac{f_3}{f_3^2+1} & 1 \end{vmatrix} = 0.$$

This acnodal nomogram is shown in Figure 3.33, rotated to fit the page.

The third form of single-scale nomogram from Clark is for the equation $f_1 f_2 f_3 = 1$ of (3.7.9). This corresponds to $A = 0$, $B = 0$ and $C = -1$ in (3.7.7). The determinant equation reduces to

$$\begin{vmatrix} \frac{f_1}{f_1^3-1} & -\frac{f_1^2}{f_1^3-1} & 1 \\ \frac{f_2}{f_2^3-1} & -\frac{f_2^2}{f_2^3-1} & 1 \\ \frac{f_3}{f_3^3-1} & -\frac{f_3^2}{f_3^3-1} & 1 \end{vmatrix} = 0,$$

and this crunodal nomogram, a very distinctive curve known as the *folium of Descartes*, is drawn in Figure 3.34.

Finally, a simple cubic curve can provide a single-scale nomogram for the equation $f_1 + f_2 + f_3 = 0$ [**259**, pp. 251–252]. Let's perform the transformation in (3.5.7) on the cuspidal determinant (3.7.10) for $\frac{1}{f_1} + \frac{1}{f_2} + \frac{1}{f_3} = 0$ to bring the points at infinity to zero and vice-versa:

$$\overline{f}_i = \frac{f_i}{g_i}, \quad \overline{g}_i = \frac{1}{g_i}.$$

We obtain the determinant equation

$$\begin{vmatrix} \frac{1}{f_1} & \frac{1}{f_1^3} & 1 \\ \frac{1}{f_2} & \frac{1}{f_2^3} & 1 \\ \frac{1}{f_3} & \frac{1}{f_3^3} & 1 \end{vmatrix} = 0.$$

If we perform variable substitutions to replace f_i and g_i with their reciprocals, we now have for the equation $f_1 + f_2 + f_3 = 0$ the determinant equation

(3.7.11)
$$\begin{vmatrix} f_1 & f_1^3 & 1 \\ f_2 & f_2^3 & 1 \\ f_3 & f_3^3 & 1 \end{vmatrix} = 0.$$

in which the scales lie on the curve $\eta = \xi^3$. This is shown in Figure 3.35, which also includes the Cartesian axes for reference. Another way of looking at this is that three roots of a third-order equation $ax^3 + bx^2 + cx + d = 0$ add to $-\frac{b}{a}$. So if we label a plot of $y = x^3$ with its value of x at every tick mark, we have a nomogram for addition with a single shared scale.

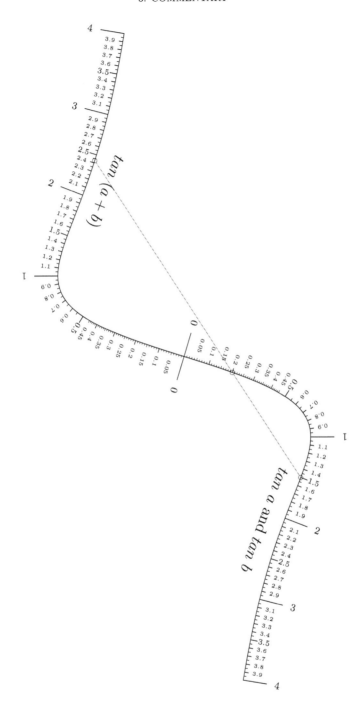

FIGURE 3.33. Acnodal nomogram (rotated) for $\tan(a+b) = \frac{\tan a + \tan b}{1 - \tan a \tan b}$.

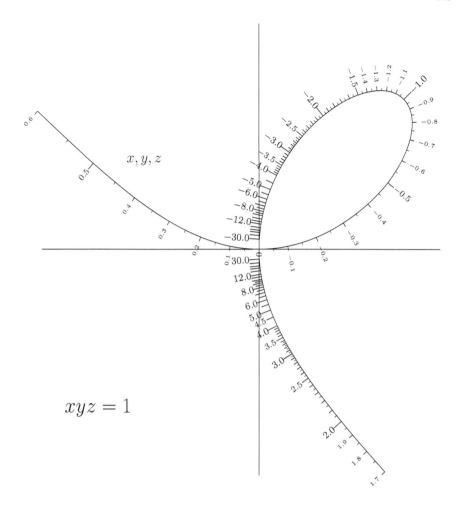

FIGURE 3.34. Crunodal nomogram for $xyz = 1$.

Further, we can perform shear transformations, one for each axis, on (3.7.11) to spread the positive ξ and η axes, as in

$$\overline{f_i} = f_i - 12f_i, \quad \overline{g_i} = g_i - 3f_i.$$

This produces the nomogram in Figure 3.36 [**259**, p. 260], one that can be quite usable when adjusted for ranges of functions of the three variables.

Clark's work was important in the history of nomography. However, as we have seen in Section 3.5, Gronwall placed this work on single-scale nomograms in the larger context of Weierstrass's elliptic functions. He also derived new families of single-scale nomograms and extended Clark's work by introducing

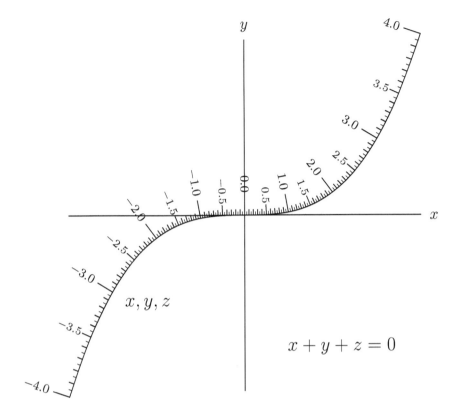

FIGURE 3.35. Cubic nomogram for $x + y + z = 0$.

the free parameters x_0, y_0 and a. It is worth noting that Gronwall was generous in acknowledging Clark's achievements, and it is apparent that he treated conical nomograms with special care in his paper.

3.8. Gronwall's Contribution to Nomography

By the time Gronwall turned his attention to matters of nomography, standard determinant equations had been derived for the most common equations in engineering. As he states at the beginning of his paper, "For specific classes of equations $[F(x, y, z) = 0]$, certainly the majority of equations encountered in practice, we know how to accomplish reduction to the [standard determinant form]." And there were wondrous designs in works by d'Ocagne and Soreau, among others, before Gronwall.

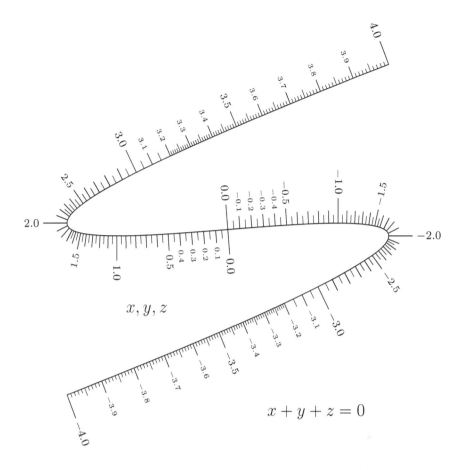

FIGURE 3.36. Cubic nomogram (sheared) for $x+y+z=0$.

But how much time was spent trying to design nomograms for equations for which it was not possible? Gronwall was concerned with this larger issue, and in fact existence theorems have occupied nomography theorists for over a century. Gronwall succeeds for the first time in providing a necessary and sufficient condition for identifying three-variable equations that can be represented by a nomogram. His solution is a test for a common integral of a system of two partial differential equations, a complicated proposition. But Gronwall's paper is more than this. His explorations of the properties of this system of equations led to significant simplifications in some important forms of nomogram. In the process he introduced new families of nomograms and new classes of equations that can be written as the simple sum of functions of the variables, including conical and single-scale nomograms. In fact, he catalogued all possible families of nomogram for this type of equation, the only one encompassing more than one. In this regard Gronwall's paper reveals an effort by an applied mathematician to generate practical results from his theoretical work. It is a remarkable achievement.

Bibliography – Thomas Hakon Gronwall

[1] Thomas Hakon Gronwall, Letters to Gösta Mittag-Leffler (1877-1932), The Royal Swedish Academy of Sciences, Djursholm, Sweden.

[2] _____, Om system af lineära totala differentialekvationer, ÖKVAF **52** (1895), 729–757.

[3] _____, Ueber Integrale algebraischer Differentialausdrucke von mehreren Veränderlichen, ÖKVAF **53** (1896), 67–72.

[4] _____, Några anvädnigar af de 2n-periodiska funktionerna på teorin für system af lineära totala differentialekvationer, ÖKVAF **53** (1896), 295–314.

[5] _____, Note sur les fonctions et les nombres algébriques, ÖKVAF **54** (1897), 199–203.

[6] _____, Deux théorèmes sur les nombres transcendants, ÖKVAF **54** (1897), 623–632.

[7] _____, Sur les nombres transcendants. II, ÖKVAF **55** (1898), 153–156.

[8] _____, Om system af linjära totala differentialekvationer sürskilt sådana med $2n$-periodiska koeificienter, Akademisk afhandling, Uppsala, 1898, 52 p.

[9] _____, Eine Verallgemeinerung der Lamé'schen Differentialgleichung, ÖKVAF **55** (1898), 357–368.

[10] _____, Sur les fonctions qui ne satisfont à aucune équation différentielle algébrique, ÖKVAF **55** (1898), 387–395.

[11] _____, Sur les singularités des systémes d'équations linéaires aux différentielles totales, BKSVAH, vol. 24 (1:9), 1899, 22 p.

[12] _____, Sur Les Équations entre Trois Variables Représentables par des Nomogrammes à Points Alignés, J. de Math. **8** (1912), 59–102.

[13] _____, On Analytic Functions of Constant Modulus on a Given Contour, Ann. Math., (2) **14** (1912), no. 1/4, 72–80.

[14] _____, On a theorem of Fejér's and an analogon to Gibbs' phenomenon, Trans. Amer. Math. Soc. **13** (1912), no. 4, 445–468, [18 (4-5-12) 445–446].

[15] _____, Über die Gibbssche Erscheinung und die trigonometrischen Summen $\sin x + 1/2 \sin 2x + \ldots + 1/n \sin nx$, Math. Ann. **72** (1912), no. 2, 228–243.

[16] _____, Über die Lebesgueschen Konstanten bei den Fourierschen Reihen, Math. Ann. **72** (1912), no. 2, 244–261.

[17] _____, Problems for Solution: Calculus: 331, Amer. Math. Monthly **19** (1912), no. 10/11, 175, Gives title as Ph.D., C. E., 912 Schiller Building, Chicago, Illinois.

[18] _____, Sur un théorème de M. Picard, C. R. Acad. Sci. Paris Sér. A-B **155** (1912), 764–7661, [19 (11-30-12) 225–226].

[19] _____, On Lebesgue's Constants in the Theory of Fourier's Series, Ann. Math., (2) **15** (1913), no. 1/4, 125–128.

[20] _____, Über die Laplacesche Reihe, Math. Ann. **74** (1913), no. 2, 213–270, [19 (9-10-12) 61-62].

[21] _____, Some Special Boundary Problems in the Theory of Harmonic Functions, Bull. Amer. Math. Soc. **19** (1913), no. 5, 227–233, [19 (9-11-12) 61].

[22] _____, Some Asymptotic Expressions in the Theory of Numbers, Trans. Amer. Math. Soc. **14** (1913), no. 1, 113–122, [19 (4-6-12) 447].

[23] _____, On the Summability of Fourier's Series, Bull. Amer. Math. Soc. **20** (1913), no. 3, 139–146, [19 (2-22-13) 392].

[24] _____, Problems and Questions: Problems for Solution: Calculus: 339, Amer. Math. Monthly **20** (1913), no. 6, 196.

[25] _____, Sur les séries de Dirichlet correspondant à des caractères complexes, Rend. Circ. Mat. Palermo **35** (1913), 145–159, [19 (11-30-12) 225].

[26] _____, Über das Verhalten der Riemannschen Zetafunktion auf der Geraden $\sigma = 1$, Archiv der Math. und Physik **20** (1913), no. 3, 231–238, [19 (11-30-12) 224].

[27] _____, Sur la fonction $\zeta(s)$ de Riemann au voisinage de $\sigma = 1$, Rend. Circ. Mat. Palermo **35** (1913), 95–102, [19 (11-30-12) 225].

[28] _____, On Weierstrass's preparation theorem, [19 (1-2-13) 298], 1913.

[29] _____, On systems of linear total differential equations, [20 (12-31-13) 299], 1913.

[30] _____, Extension of Laurent's theorem to several variables, [20 (12-31-13) 299], 1913.

[31] _____, Some Remarks on Conformal Representation, Ann. Math., (2) **16** (1914), no. 1/4, 72–76, 138, [21 (10-31-14) 165].

[32] _____, On the Degree of convergence of Laplace's Series, Trans. Amer. Math. Soc. **15** (1914), no. 1, 1–30, [19 (2-22-13) 392].

[33] _____, On the Maximum Modulus of an Analytic Function, Ann. Math., (2) **16** (1914), no. 1/4, 77–81, [19 (4-26-13) 507].

[34] _____, An Integral Equation of the Volterra Type, Ann. Math., (2) **16** (1914), no. 1/4, 119–122, [21 (1-2-15) 284].

[35] _____, Book Review: *Grundlagen der Geometrie* by David Hilbert, Bull. Amer. Math. Soc. **20** (1914), no. 6, 325.

[36] _____, Book Review: *Grundzüge einer allgemeinen Theorie der linearen Integralgleichungen* by David Hilbert, Bull. Amer. Math. Soc. **20** (1914), no. 6, 326.

[37] _____, Book Review: *Handbuch der Lehre von der Verteilung der Primzahlen* by Dr. Edmund Landau, Bull. Amer. Math. Soc. **20** (1914), no. 7, 368–376.

[38] _____, Book Review: *Transcendenz von e und π. Ein Beitrag zur höheren Mathematik vom elementaren Standpunkte aus* by Gerhard Hessenberg, Bull. Amer. Math. Soc. **20** (1914), no. 8, 421–422.

[39] _____, Book Review: *Vorlesungen über Variationsrechnung* by Oskar Bolza, Bull. Amer. Math. Soc. **20** (1914), no. 8, 422–423.

[40] _____, On Approximation by Trigonometric Sums, Bull. Amer. Math. Soc. **21** (1914), no. 1, 9–14, [20 (12-31-13) 299].

[41] _____, Book Review: *Beiträge zu einzelnen Fragen der höheren Potentialtheorie* by E. R. Neumann, Bull. Amer. Math. Soc. **21** (1914), no. 2, 101–102.

[42] _____, On Lebesgue's constants in the theory of Fourier series, Ann. Math., (2) **15** (1914), no. 2, 125–128, [19 (2-22-13) 392].

[43] _____, Sur la série de Laplace, C. R. Acad. Sci. Paris Sér. A-B **158** (1914), 1488–1490, [19 (1-2-13) 299].

[44] _____, Sur quelques méthodes de sommation et leur application à la série de Fourier, C. R. Acad. Sci. Paris Sér. A-B **158** (1914), 1664–1665, [19 (3-21-13) 442].

[45] _____, Über die Summierbarkeit der Reihen von Laplace und Legendre, Math. Ann. **75** (1914), 321–375, [19 (1-2-13) 299].

[46] _____, A Functional Equation in the Kinetic Theory of Gases, Ann. Math., (2) **17** (1915), no. 1, 1–4.

[47] _____, Book Review: *Théorie des Nombres* by E. Cahen, Bull. Amer. Math. Soc. **21** (1915), no. 6, 311–312.

[48] _____, Book Review: *Leçons sur la Théorie générale des Surfaces et les Applications géométriques du Calcul infinitésimal* by Gaston Darboux, Bull. Amer. Math. Soc. **21** (1915), no. 6, 312–313.

[49] _____, Book Review: *Die Berührungstransformationen: Geschichte und Invariantentheorie* by H. Liebmann and F. Engel, Bull. Amer. Math. Soc. **21** (1915), no. 7, 357–358.

[50] _____, Book Review: *Elementarmathematik vom höheren Standpunkte aus*. Teil II: *Geometrie* by F. Klein, Bull. Amer. Math. Soc. **21** (1915), no. 7, 360.

[51] _____, Book Review: *Elementare Algebra* by Eugen Netto, Bull. Amer. Math. Soc. **21** (1915), no. 7, 360.

[52] _____, Book Review: *Einführung in die Vektoranalysis mit Anwendungen auf die mathematische Physik* by Richard Gans, Bull. Amer. Math. Soc. **21** (1915), no. 7, 360–361.

[53] _____, Book Review: *Darstellende Geometrie des Geländes* by Richard Gans, Bull. Amer. Math. Soc. **21** (1915), no. 7, 361.

[54] _____, Sur les surfaces minima formant une famille de Lamé, C. R. Acad. Sci. Paris Sér. A-B **161** (1915), 582–583, [20 (4-25-14) 517].

[55] _____, On the summation method of de la Vallee Poussin, [21 (1-2-15) 284], 1915.

[56] _____, Sur les zéros des fonctions $P(z)$ et $Q(z)$ associées à la fonction gamma, Ann. Sci. École Norm. Sup. (3) **33** (1916), 381–393, [22 (2-26-16) 375].

[57] _____, A Problem in Geometry Connected with the Analytic Continuation of a Power Series, Ann. Math., (2) **18** (1916), no. 2, 65–69, [23 (9-5-16) 81].

[58] _____, On the Power Series for $\log(1 + z)$, Ann. Math., (2) **18** (1916), no. 2, 70–73, [23 (9-5-16) 80].

[59] _____, On the Convergence of Binet's Factorial Series for $\log \Gamma(s)$ and $\psi(s)$, Ann. Math., (2) **18** (1916), no. 2, 74–78, [23 (9-5-16) 81].

[60] _____, Book Review: Mathematische Abhandlungen, Hermann Amandus Schwarz zu seinem fünfzigjährigen Doktorjubiläum am 6. August 1914 gewidmet von Freunden und Schülern, Bull. Amer. Math. Soc. **22** (1916), no. 8, 406–407.

[61] _____, Book Review: *Graphische Methoden* by C. Runge, Bull. Amer. Math. Soc. **22** (1916), no. 8, 407–408.

[62] _____, Book Review: *Über die Theorie des Kreisels* by F. Klein and A. Sommerfeld, Bull. Amer. Math. Soc. **22** (1916), no. 8, 408.

[63] _____, Book Review: *Introduction géométrique à quelques Théories Physiques* by Émile Borel, Bull. Amer. Math. Soc. **22** (1916), no. 8, 409–410.

[64] _____, Book Review: *Grundzüge der Geodäsie* by M. Näbauer, Bull. Amer. Math. Soc. **22** (1916), no. 8, 410–411.

[65] _____, Problems and Questions: Problems for Solution: Algebra: 469, Amer. Math. Monthly **23** (1916), no. 9, 341.

[66] _____, Sur la déformation dans la représentation conforme, C. R. Acad. Sci. Paris Sér. A-B **162** (1916), 249–252, [21 (1-2-15) 285; 22 (2-26-16) 376.].

[67] _____, Sur la déformation dans la représentation conforme sous des conditions restrictives, C. R. Acad. Sci. Paris Sér. A-B **162** (1916), 316–318, [22 (2-26-16) 376].

[68] _____, A functional equation in the kinetic theory of gases, Ann. Math., (2) **17** (1916), 1–4, [21 (2-24-15) 484].

[69] _____, Determination of all triply orthogonal systems containing a family of minimal surfaces, Ann. Math., (2) **17** (1916), 76–100, [20 (4-25-14) 517].

[70] _____, An elementary exposition of the theory of the gamma function, Ann. Math., (2) **17** (1916), 124–166, Annotated translation of a Danish memoir by J. L. W. V. Jensen.

[71] _____, Sur une équation fonctionnelle dans la théorie cinétique des gaz, C. R. Acad. Sci. Paris Sér. A-B **162** (1916), 415–418, [22 (2-26-16) 375].

[72] _____, Über einige Summationsmethoden und ihre Anwendung auf die Fouriersche Reihe, J. reine angew. Math. **147** (1917), 16–35, [19 (3-21-17) 442].

[73] _____, On the expressibility of a uniform function of several complex variables as the quotient of two functions of entire character, Trans. Amer. Math. Soc. **18** (1917), no. 1, 50–64, [20 (10-25-13) 173-174].

[74] _____, Elastic Stresses in an Infinite Solid with a Spherical Cavity, Ann. Math., (2) **19** (1918), no. 4, 295–296, [22 (2-26-16) 379].

[75] _____, The Gamma Function in the Integral Calculus, Ann. Math., (2) **20** (1918), no. 2, 35–124.

[76] _____, Equipotential minimal surfaces, [25 (12-28-18) 256], 1918.

[77] _____, On Kummer's series, [25 (12-28-18) 256], 1918.

[78] _____, A Small Arc Method For Computing Trajectories. Preliminary Communication, APG-6 32, Aberdeen Proving Grounds, Aberdeen, Maryland, 17 December 1918, Three pages.

[79] _____, On the Influence of Keyways on the Stress Distribution in Cylindrical Shafts, Trans. Amer. Math. Soc. **20** (1919), no. 3, 234–244, [22 (2-26-16) 379].

[80] _____, On the Shortest Line Between Two Points in Non-Euclidean Geometry, Ann. Math., (2) **20** (1919), no. 3, 200–201.

[81] _____, On a System of Linear Partial Differential Equations of the Hyperbolic Type, Ann. Math., (2) **20** (1919), no. 4, 274–278.

[82] _____, Note on the Derivatives with Respect to a Parameter of the Solutions of a System of Differential Equations, Ann. Math., (2) **20** (1919), no. 4, 292–296.

[83] _____, Problems and Solutions: Problems for Solution: 2784, Amer. Math. Monthly **26** (1919), no. 8, 366.

[84] _____, Problems and Solutions: Problems for Solution: Calculus: 339, Amer. Math. Monthly **26** (1919), no. 5, 213.

[85] _____, Nomogram for Computing Range and Deflection from Co-Ordinates, June 2, 1919 1919, Records of the Department of Ordnance, Record Group 156, Entry 866, National Archives and Record Administration.

[86] _____, On the Computation of Differential Corrections, January 24, 1919 1919, Records of the Department of Ordnance, Record Group 156, Entry 866, National Archives and Record Administration.

[87] _____, On the Twist in Conformed Mapping, Proc. Nat. Acad. Sci. USA **5** (1919), no. 7, 248–250, Range Firing Section, Aberdeen Proving Ground; Communicated by E. H. Moore, April 29, 1919; Note II on Conformal Mapping under aid of Grant No. 207 from the Bache Fund.

[88] _____, A Theorem on Power Series, with an Application to Conformal Mapping, Proc. Nat. Acad. Sci. USA **5** (1919), no. 7, 22–24, [25 (12-28-18) 256].

[89] _____, Investigation of a class of fundamental inequalities in the theory of analytic functions, Ann. Math., (2) **21** (1919), 1–29, Annotated translation of a Danish paper by J. L. W. V. Jensen.

[90] _____, A Theorem on Power Series, with an Application to Conformal Mapping, Proc. Nat. Acad. Sci. USA **5** (1919), no. 1, 22–24, Range Firing Section, Aberdeen Proving Ground; Communicated by E. H. Moore, December 2, 1918; Note I on Conformal Mapping under aid of Grant No. 207 from the Bache Fund.

[91] _____, On the Computation of Differential Corrections, D-I-2 18, Aberdeen Proving Grounds, Aberdeen, Maryland, 14 January 1919, Seven pages.

[92] _____, Qualitative Properties of the Ballistic Trajectory, A-I-G 32, Aberdeen Proving Grounds, Aberdeen, Maryland, 20 February 1919, Eight pages.

[93] _____, Nomogram For Computing Range and Deflection From Coordinates, E-iii-6, Aberdeen Proving Grounds, Aberdeen, Maryland, 1919, Four pages.

[94] _____, On the Construction of Maps For the Determination of Range and Deflection From Observed Azimuths, E-iii-7 81, Aberdeen Proving Grounds, Aberdeen, Maryland, 8 July 1919, Fifteen pages.

[95] _____, Computation of Deflection For Target Records, E-iii-9 82, Aberdeen Proving Grounds, Aberdeen, Maryland, 10 July 1919, One page.

[96] _____, On Differential Corrections and Weighting Factors, Tech. Report T. S. 101, Office of Ballistics, Washington, D. C., August 1919, Twenty-two pages.

[97] ———, Qualitative Properties of the Ballistic Trajectory, Ann. Math., (2) **22** (1920), no. 1, 44–65, [25 (4-26-19) 441; 26 (2-28-20) 340].

[98] ———, Conformal Mapping of a Family of Real Conics upon Another, Ann. Math., (2) **22** (1920), no. 2, 101–127.

[99] ———, On The Distortion in Conformal Mapping when the Second Coefficient in the Mapping Function has an Assigned Value, Proc. Nat. Acad. Sci. USA **6** (1920), no. 6, 300–302, [26 (2-28-20) 340]. Technical Staff, Office of The Chief of Ordnance, Washington, D. C.; Communicated by E. H. Moore, April 27, 1920; Note III On Conformal Mapping Under Aid of Grant No. 207 From the Bache Fund.

[100] ———, Conformal Mapping of a Family of Real Conics upon Another, Proc. Nat. Acad. Sci. USA **6** (1920), no. 6, 312–315.

[101] ———, Upper bounds for the coefficients in conformal mapping, [27 (9-8-20) 65], 1920.

[102] ———, Some inequalities in the theory of functions of a complex variable, [27 (10-30-20) 103], 1920.

[103] ———, Range Variations Due to Changes in Gravity, Tech. Report T. S. 122, Office of Ballistics, Washington, D. C., January 1920, Two pages.

[104] ———, Differential Variations in Ballistics, with Applications to the Qualitative Properties of the Trajectory, Trans. Amer. Math. Soc. **22** (1921), no. 4, 505–525, [26 (10-25-19) 149].

[105] ———, A Sequence of Polynomials Connected with the nth Roots of Unity, Bull. Amer. Math. Soc. **27** (1921), no. 6, 275–279, [27 (9-8-20) 64-65].

[106] ———, On the Fourier Coefficients of a Continuous Function, Bull. Amer. Math. Soc. **27** (1921), no. 7, 320–321, [27 (9-8-20) 64].

[107] ———, Some empirical formulas in ballistics, [27 (2-26-21) 349], 1921.

[108] ———, On biharmonic functions, [28 (10-29-21) 93], 1921.

[109] ———, Approximate Formulas for the Variations in Range and Deflection due to the Rotation of the Earth, Tech. Report T. S. 162, Office of Ballistics, Washington, D. C., April 1921, Four pages.

[110] ———, T. S. 240, Technical Staff of the Office of Ballistics, Washington, D. C., 1921, Missing from archive.

[111] ———, Angle of max. range of 14″ Naval Gun MV = 914 meters = 3000 ft., C = 10.5, Tech. Report T. S. 242, Office of Ballistics, Washington, D. C., 21 June 1921.

[112] ———, Cross Wind Weighting Factors, Tech. Report T. S. 243, Office of Ballistics, Washington, D. C., 15 October 1921.

[113] ———, Summation of a Double Series, Ann. Math., (2) **23** (1922), no. 3, 282–285, [27(2-26-21) 349].

[114] ———, On Power Series with Positive Real Part in the Unit Circle, Ann. Math., (2) **23** (1922), no. 4, 317–332, [28 (9-8-21) 9-10].

[115] ———, Qualitative properties of the ballistic trajectory (second paper), [28 (2-25-22) 239], 1922.

[116] ———, Qualitative Properties of the Ballistic Trajectory. Second Paper, Tech. Report T. S. 241, Office of Ballistics, Washington, D. C., March 1922.

[117] ———, Differential variations in Range to any point on a Trajectory, Tech. Report T. S. 244, Office of Ballistics, Washington, D. C., March 1922.

[118] ———, Interpolation to halves. Body of the Table: Last two digits of the smaller interpolated difference, Tech. Report T. S. 245, Office of Ballistics, Washington, D. C., 6 March 1922.

[119] ———, Yards to meters. Body of the table gives the smallest number of yds. that will give the argument, Tech. Report T. S. 246, Office of Ballistics, Washington, D. C., June 1922.

[120] ———, Traverse tables, Tech. Report T. S. 247, Office of Ballistics, Washington, D. C.

[121] ———, Density tables, Tech. Report T. S. 248, Office of Ballistics, Washington, D. C., 14 July 1922.

[122] _____, Firing Tables for gun mounted on an aircraft, Tech. Report T. S. 249, Office of Ballistics, Washington, D. C., September 1922.

[123] _____, Isothermal surfaces with spherical lines of curvature in one system, [29 (4-28-23) 218], 1923.

[124] _____, Extension of Tchebychef's statistical theorem, [29 (4-28-23) 218], 1923.

[125] _____, Extension of a theorem due to Lerch, [31 (10-25-24) 109], 1924.

[126] _____, The Mutual Inductance of Two Square Coils, Trans. Amer. Math. Soc. **27** (1925), no. 4, 516–536, [29 (4-28-23) 208].

[127] _____, The behavior at infinity of the gamma and associated functions, [31 (5-2-25) 484], 1925.

[128] _____, The algebraic structure of the formulas in plane trigonometry. Three papers, [31 (5-2-25) 484; 32 (9-11-25) 23; 32 (10-31-25) 36], 1925.

[129] _____, On the Existence and Properties of the Solutions of a Certain Differential Equation of the Second Order, Ann. Math., (2) **28** (1926), no. 1/4, 355–364.

[130] _____, Errata: "On the zeros of the function $\beta(z)$ associated with the gamma function" [Trans. Amer. Math. Soc. **28** (1926), no. 3, 391–399; 1501352], Trans. Amer. Math. Soc. **28** (1926), no. 4, 786.

[131] _____, On the zeros of the function $\beta(z)$ associated with the gamma function, Trans. Amer. Math. Soc. **28** (1926), no. 3, 391–399, 786, [23 (9-5-16) 81].

[132] _____, Book Review: *Sieben-und mehrstellige Tafeln der Kreis-und Hyperbelfunktionen und deren Produkte sowie der Gammafunktion, nebst einem Anhang: Interpolations-und sonstige Formeln* by Keiichi Hayashi, Bull. Amer. Math. Soc. **32** (1926), no. 6, 716.

[133] Thomas Hakon Gronwall and V. K. La Mer, On extensions of the Debye–Hückel theory of strong electrolytes to concentrated solution, Science **64** (1926), 122, [32 (5-1-26) 316].

[134] Thomas Hakon Gronwall, Reflection of Radiation from a Finite Number of Equally Spaced Parallel Planes, Phys. Rev. **27** (1926), no. 3, 277–285, [28 (2-25-22) 239].

[135] _____, Book Review: *Traité de Balistique Extérieure* by P. Charbonnier, Bull. Amer. Math. Soc. **33** (1927), no. 1, 120.

[136] _____, On the Determination of the Apparent Diameter of the Ions in the Debye–Hückel Theory of Strong Electrolytes, Proc. Nat. Acad. Sci. USA **13** (1927), no. 4, 198–202, Department of Chemistry, Columbia University, Contribution of the Department of Chemistry, Columbia University, No. 538.

[137] _____, The Longitudinal Vibrations of a Liquid Contained in a Tube with Elastic Walls, Phys. Rev. **30** (1927), no. 1, 71–83, [33 (2-26-27) 258].

[138] Thomas Hakon Gronwall and V. K. LaMer, The Partial Molal Volumes of Water and Salt in Solutions of the Alkali Halides, J. Phys. Chem. **31** (1927), no. 3, 393–406.

[139] Thomas Hakon Gronwall, A new form of the remainder in the binomial series with applications, [32 (2-27-26) 195], 1927.

[140] _____, Book Review: *Die Gewöhnlichen und Partiellen Differenzengleichungen der Baustatik.* by F. Bleich and E. Melan, Bull. Amer. Math. Soc. **34** (1928), no. 6, 787.

[141] _____, Book Review: *Formules Stokiennes* by A. Buhl, Bull. Amer. Math. Soc. **34** (1928), no. 6, 790.

[142] Thomas Hakon Gronwall, V. K. LaMer, and K. Sandved, Über den Einfluss der Sogenannten Höhere Glieder in der Debye Hückelschen Theorie der Lösung Starker Electrolyte, Phys. Z. **29** (1928), 358–393.

[143] Thomas Hakon Gronwall, On the convergence region of a power series in several variables, [34 (4-6-28) 423], 1928.

[144] _____, Questions and Discussions: Discussions: The Number of Arithmetical Operations Involved in the Solution of a System of Linear Equations, Amer. Math. Monthly **36** (1929), no. 6, 325–327.

[145] _____, Book Review: *Lehrbuch der Kombinatorik* by Eugen Netto with Viggo Brun and Th. Skolem, Bull. Amer. Math. Soc. **35** (1929), no. 4, 577–578.

[146] _____, Book Review: *Vorlesungen über Algebra* by L. Bieberbach and Dr. Gustav Bauer, Bull. Amer. Math. Soc. **35** (1929), no. 4, 581.
[147] _____, Book Review: *Leçons sur quelques Types Simples d'Équations aux Dérivées Partielles avec des Applications à la Physique Mathématique* by Émile Picard, Bull. Amer. Math. Soc. **35** (1929), no. 5, 733–734.
[148] _____, Book Review: *The new Quantum Mechanics* by G. Birtwistle, Bull. Amer. Math. Soc. **35** (1929), no. 5, 736.
[149] _____, Book Review: *Traité d'Analyse* by Émile Picard, Bull. Amer. Math. Soc. **35** (1929), no. 5, 736.
[150] _____, On the Number of Arithmetical Operations Involved in the Solution of a System of Linear Equations, Amer. Math. Monthly **36** (1929), 235–237, [35 (3-30-29) 443].
[151] _____, On the differential equation of the vibrating membrane, [35 (2-23-29) 293], 1929.
[152] _____, Straight line geodesics in Einstein's parallelism geometry. Two papers, [35 (2-23-29) 293; 35 (3-29-29) 437], 1929.
[153] _____, Rotational symmetry in the characteristic functions of the Schrödinger wave equation, [36 (2-22-30) 218], 1929.
[154] _____, On Minkowski's Mixed Volume of Three Convex Solids, Ann. Math., (2) **31** (1930), no. 3, 470–472, [27 (6-17-20) 1, title but no abstract].
[155] _____, The Converse of Euler's Theorem on Homogeneous Functions, Ann. Math., (2) **31** (1930), no. 3, 473–474, [33 (2-26-27) 258-259].
[156] _____, A Formula in Geometrical Optics, Ann. Math., (2) **31** (1930), no. 3, 475–478, [36 (10-26-29) 51].
[157] _____, Zur Gibbsschen Erscheinung, Ann. Math., (2) **31** (1930), no. 2, 233–240, [31 (2-28-25) 302; 32 (5-1-26) 316].
[158] _____, Book Review: *Operational Circuit Analysis* by Vannevar Bush, Bull. Amer. Math. Soc. **36** (1930), no. 1, 37–38.
[159] _____, Book Review: *Mathematical and Physical Papers* by Joseph Larmor, Bull. Amer. Math. Soc. **36** (1930), no. 7, 470–471.
[160] _____, Book Review: *Analytische Geometrie* by L. Bieberbach, Bull. Amer. Math. Soc. **36** (1930), no. 7, 471–472.
[161] _____, Book Review: *Die Lehre von den Kettenbrüchen* by O. Perron, Bull. Amer. Math. Soc. **36** (1930), no. 9, 615.
[162] _____, Book Review: *Les Problèmes des Isopérimètres et des Isépiphanes* by Tommy Bonnesen, Bull. Amer. Math. Soc. **36** (1930), no. 9, 617.
[163] _____, A Diophantine Equation Connected with the Hydrogen Spectrum, Phys. Rev. **36** (1930), no. 11, 1671–1672, [33 (10-30-26) 12].
[164] Thomas Hakon Gronwall and V.K. La Mer, The variation of the dielectric constant in the Debye–Hückel theory., [33 (10-30-26) 12], 1930.
[165] Thomas Hakon Gronwall, On the Wave Equation of the Hydrogen Atom, Ann. Math., (2) **32** (1931), no. 1, 47–52, [34 (9-6-28) 703].
[166] _____, On the Cèsaro Sums of Fourier's and Laplace's Series, Ann. Math., (2) **32** (1931), no. 1, 53–59, [31 (10-25-24) 109-110].
[167] _____, A Functional Equation in Differential Geometry, Ann. Math., (2) **32** (1931), no. 2, 313–326, [25 (12-28-18) 255].
[168] _____, Über den Konvergenzbereich der Potenzreihenentwicklung einer harmonischen Funktion von n Veränderlichen, Math. Zeitschrift **33** (1931), no. 1, 177–185, [32 (1-2-26) 127].
[169] Thomas Hakon Gronwall, V. K. LaMer, and Lotti J. Greiff, Theory in the Case of Unsymmetric Valence Type Electrolytes, J. Phys. Chem. **35** (1931), no. 8, 2245–2288, Hille has the title as: The influence of the higher terms of the Debye–Hückel theory in the case of unsymmetric valence type electrolytes.

[170] _____, Errata. Theory in the Case of Unsymmetric Valence Type Electrolytes, J. Phys. Chem. **35** (1931), no. 10, 3103–3104.
[171] Thomas Hakon Gronwall, A new form of the remainder in the binomial series with applications, [32 (10-31-25) 36-37], 1931.
[172] Thomas Hakon Gronwall, V. K. LaMer, and Lotti J. Greiff, Errata. Theory in the Case of Unsymmetric Valence Types of Electrolytes, J. Phys. Chem. **35** (1931), no. 12, 3692–3692.
[173] Thomas Hakon Gronwall, On the theory of potentiometric titration, [37 (9-8-31) 530], 1931.
[174] _____, Summation of Series and Conformal Mapping, Ann. Math., (2) **33** (1932), no. 1, 101–117, [32 (10-31-25) 36].
[175] _____, An Inequality for the Bessel Functions of the First Kind with Imaginary Argument, Ann. Math., (2) **33** (1932), no. 2, 275–278, [37 (9-8-31) 530].
[176] _____, A Special Conformally Euclidean Space of Three Dimensions Occurring in Wave Mechanics, Ann. Math., (2) **33** (1932), no. 2, 279–293, [37 (9-8-31) 530].
[177] _____, The Helium Wave Equation, Phys. Rev. **51** (1937), no. 8, 655–660.

References

[178] M. d'Ocagne, Traité de Nomographie, Gauthier-Villars, Paris, 1899.
[179] ———, Traité de Nomographie, 2 ed., Gauthier-Villars, Paris, 1921.
[180] Book Review: A Course of Instruction in Ordnance and Gunnery by Henry Metcalfe, Journal of the U S. Artillery **1** (1892), 74–76.
[181] H. T. Eddy, Modern Graphical Developments, Mathematical papers read at the International Mathematics Congress: held in connection with the World Columbian Exposition, Chicago, 1893, Macmillan, Chicago, 1896, pp. 58–71.
[182] F. Morley, Book Review: *Traité de Nomographie* by M. d'Ocagne, Bulletin of the American Mathematical Society **6** (1900), 398–400.
[183] M. J. Eichhorn, The Construction and Use of Graphical Tables, Western Electrician (1901), 162–163.
[184] Book Review: *Manuale del Tiro, By Commandante G. Ronca, with an appendix on "Nomography"* by Prof. Pesci, Journal of the U. S. Artillery **19** (1903), 220–221.
[185] E. Kasner, The Present Problems of Geometry, Bulletin of the American Mathematical Society **11** (1905), 283–314.
[186] E. H. Moore, The Cross-Section Paper as a Mathematical Instrument, The School Review **14** (1906), no. 5, 317–338.
[187] J. Clark, Théorie générale des abaques d'alignment de tout ordre, Revue de Mécanique **21** (1907), 321–335.
[188] ———, Théorie générale des abaques d'alignment de tout ordre, Revue de Mécanique **21** (1907), 575–585.
[189] R. de Beaurepaire, Graphs and Abacuses, Madras, 1907.
[190] O. Lissak, Ordnance and Gunnery, John Wiley and Sons, New York, 1907.
[191] J. Clark, Théorie générale des abaques d'alignment de tout ordre, Revue de Mécanique **22** (1908), 236–253.
[192] ———, Théorie générale des abaques d'alignment de tout ordre, Revue de Mécanique **22** (1908), 451–472.
[193] R. de Aquino, Nomograms for Deducing Altitude and Azimuth and for Star Identification and Finding Course and Distance in Great Circle Sailing, U. S. Naval Institute, Annapolis, 1908.
[194] L. Hewes, Book Review: *Calcul Graphique et Nomographie* and *Le Calcul Simplifié* by M. d'Ocagne, Bulletin of the American Mathematical Society **15** (1908), 127–132.
[195] J. Peddle, The Construction of Graphical Charts, The American Machinist (1908).
[196] L. Hewes, A Graphic Solution of Kutter's Formula, Transactions of the American Institute of Mining Engineers, New Haven Meeting (1909), 231–232.
[197] J. Peddle, The Construction of Graphical Charts, McGraw-Hill, New York, 1910.
[198] C. Runge, Columbia University Lectures: Graphical Methods, The New Era Printing Company, Lancaster, 1910.
[199] Department of Commerce and Bureau of the Census Labor, Thirteenth Census of the United States: 1910 Population, U.S. Government Printing Office, 1910.

[200] M. J. Eichhorn, Nomogram for the Properties of Steam, The Swedish Engineer's Society of Chicago **3** (1911), 1–8.
[201] C. Runge, Graphical Methods, New York, Columbia University Press, 1912.
[202] F. A. Halsey, Handbook for Machine Designers and Draftsmen, McGraw-Hill, New York, 1913.
[203] R. K. Hezlet, Nomography or the Graphic Representation of Formula, The Royal Artillery Institution, Woolwich, 1913.
[204] E. J. Wilcynski, A Forgotten Theorem of Newton's on Planetary Motion and An Instrumental Solution of Kepler's Problem, The Astronomical Journal **27** (1913), no. 644, 155–156.
[205] Faculty Minutes, Report of the Committee of the Curriculum, September 24, 1914, Seeley G. Mudd Manuscript Library, Princeton, 1914.
[206] W. C. Brinton, Graphic Methods for Presenting Facts, The Engineering Magazine Company, New York, 1914.
[207] E. V. Huntington, The Faultless Faultfinder, The Engineering and Mining Journal **98** (1914), 291–296.
[208] The Princeton University Annual Reports of the President and the Treasurer for the Year Ending December, 1914, 1914.
[209] O. D. Kellogg, Nomograms with points in alignment, Zeitschrift für Mathematik und Physik **63** (1915), 159–173.
[210] R. C. Strachan, Nomographic Solutions for Formulas of Various Types, Transactions of the American Society of Civil Engineers (1915), Paper No. 1333.
[211] L. S. Marks, The Mechanical Engineers' Handbook, McGraw-Hill, New York, 1916.
[212] L. Hewes, Nomograms of Adjustment, Annals of Mathematics **18** (1917), no. 4, 194–199.
[213] J. Lipka, A Manual of Mathematics, John Wiley and Sons, New York, 1917.
[214] H. G. Deming, A Manual of Chemical Nomography, The University of Illinois Press, Champaign, 1918.
[215] E. V. Huntington, Handbook for Mathematics for Engineers, McGraw-Hill, New York, 1918.
[216] J. Lipka, Graphical and Mechanical Computation, John Wiley and Sons, New York, 1918.
[217] G. Wentorth, D. E. Smith, and W. Schlaugh, Commercial Algebra, Book II, Ginn and Company, New York, 1918.
[218] N. T. M. Wilsmore, Book Review: *A Manual of Chemical Nomography*, Analyst (1918), 394–396.
[219] C. S. Aitchison, Type Formulae for Nomograms, Master's thesis, University of Washington, 1919, Submitted 1928.
[220] Book Review: *Graphical and Mechanical Computation* by J. Lipka, American Mathematical Monthly **26** (1919), no. 5, 203.
[221] A. A. Bennett, Book Review: *Elementary Mathematics for Field Artillery*, American Mathematical Monthly **26** (1919), no. 8, 353–355.
[222] L. J. Ford, Elementary Mathematics for Field Artillery, Prepared and Published by Direction of the Chief of Field Artillery, 1919.
[223] F. R. Moulton, Curves of Constant Fx' and Fy', 15 March 1919.
[224] J. Peddle, The Construction of Graphical Charts, 2 ed., McGraw-Hill, New York, 1919, Revised and Enlarged.
[225] O. Veblen, Ballistic Theories and the Preparation of Range Tables, Record Group (1919), 156–866, Document 54, National Archives and Record Administration, College Park, Maryland.
[226] F. R. Moulton, History of the Ballistics Branch of the Artillery Ammunition Section, Engineering Division of the Ordnance Department for the Period April 6, 1918 to April 2, 1919, U. S. Army Military History Institute, Carlisle, 1919.

REFERENCES 159

[227] S. Brodetsky, A First Course in Nomography, G. Bell and Sons, London, 1920.
[228] E. J. Haskell, How to make and Use Graphic Charts, Codex, New York, 1920.
[229] E. Oberg and F. Jones, Shop Mathematics, The Industrial Press, New York, 1920.
[230] M. d'Ocagne, Traité de Nomographie, 2 ed., Gauthier-Villars, Paris, 1921.
[231] J. Lipka, Alignment Charts, The Mathematics Teacher **14** (1921), no. 4, 171–178.
[232] W. C. Marshall, Graphical Methods for Schools, Colleges, Statisticians, Engineers and Executives, McGraw-Hill, New York, 1921.
[233] R. Soreau, Nomographie ou Traité des Abaques, Chiron, Paris, 1921.
[234] _____, Nomographie ou Traité des Abaques, Tome II, Chiron, Paris, 1921.
[235] C. E. Hall, Nomography: As Applied to Nomograms, Master's thesis, Columbia University, New York, 1922.
[236] Book Review: *Graphical and Mechanical Computation* by J. Lipka, Coast Artillery Journal **57** (1922), no. 1, 585–586.
[237] L. Hewes and H. Seward, The Design of Diagrams For Engineering Formulas and The Theory of Nomography, McGraw-Hill, New York, 1923.
[238] J. J. Johnson, An Introduction to the Construction of Nomographic or Alignment Charts, The Coast Artillery Journal **59** (1923), no. 2, 157–172.
[239] J. Lipka, Alignment Charts for Engineers, John Wiley and Sons, New York, 1924.
[240] C. L. E. Moore, Joseph Lipka: 1883–1924, The Technology Review (1924), 258.
[241] N. Wiener, In Memory of Joseph Lipka, Journal of Mathematics and Physics (1924), no. 3, 63–65.
[242] K. G. Karsten, Charts and Graphs, Prentice-Hall, New York, 1925.
[243] H. F. MacNeish, A nomogram in n-dimensional space for the solution of n linear simultaneous equations, Bulletin of the American Mathematical Society **31** (1925), 212.
[244] O. Veblen, Letter from Oswald Veblen to Gilbert Ames Bliss, Oswald Veblen Papers, Library of Congress, Washington, D. C, 14 October 1926.
[245] E. B. Wilson, Book Review: *Design of Diagrams for Engineering Formulas and the Theory of Nomography*, Bulletin of the American Mathematical Society **32** (1926), no. 3, 295.
[246] E. Hille, Thomas Hakon Gronwall – In Memoriam, Bulletin of the American Mathematical Society **38** (1932), no. 11, 775–786.
[247] A. A. Bennett, E. W. Milne, and H. Bateman, Numerical Integration of Differential Equations, National Research Council of the National Academy of Sciences, Washington, D. C, 1933.
[248] J. Griffith, Bulletin Series No. 12: Mathematics of Alignment Chart Construction Without Use of Determinants, Engineering Experiment Station, Oregon State System of Higher Education, Oregon State College, Corvallis, 1939.
[249] H. F. MacNeish, Elementary mathematical theory of exterior ballistics, Brooklyn College Press, Brooklyn, 1942.
[250] Jahrbuch für Mathematik, 1943, JFM43.0159.03.
[251] L. J. Ford, Notre Dame Mathematical Lectures, No. 4: Alignment Charts, Notre Dame University Press, South Bend, 1944.
[252] R. D. Douglass and D. P. Adams, Elements of Nomography, McGraw-Hill, New York, 1947.
[253] D. P. Adams and H. T. Evans, Developments in the Useful Circular Nomogram, Rev. Sci. Instrum. **20** (1949), no. 3, 150–155.
[254] R. C. Fry, The Use of the Nomogram in American Industry, Master's thesis, Massachusetts Institute of Technology, Cambridge, 1950.
[255] M. W. Pentkowski, Nomographie, Akademie-Verlag, Berlin, 1953, translated by M. Peschel.
[256] _____, Skeleton Nomogram Equations of the Third Nomogrammic Order (in Russian), USSR Academy of Sciences, Moscow, 1953.

REFERENCES

[257] E. Varnum, The Value of the Circular Nomogram to an Industrial Firm, J. Eng. Drawing **19** (1955), no. 3, 36–37.

[258] L. Epstein, Nomography, Interscience Publishers, New York, 1958.

[259] E. Otto, Nomography, MacMillan, New York, 1963, Translated by J. Smólska.

[260] R. Wyss, Untersuchungen zur theoretischen Nomographie, Juris Druck + Verlag, Zürich, 1971.

[261] E. Hewitt and R. Hewitt, The Gibbs-Wilbraham phenomenon – an episode in Fourier analysis, Archive for the History of Exact Science **21** (1979), no. 2, 128–160.

[262] K. Parshall and D. Rowe, The Emergence of the American Mathematical Research Community, 1876–1900: J. J. Sylvester, Felix Klein, and E. H. Moore, American Mathematical Society, Providence, 1994.

[263] T. L. Hankins, Blood, Dirt, and Nomograms: A Particular History of Graphs, Isis **90** (1999), 50–80.

[264] A. Gluchoff, Pure Mathematics applied in early twentieth-century America: The case of T. H. Gronwall, consulting mathematician, Historia Mathematica **32** (2005), 312–357.

[265] F. Klein, Elementary Mathematics From An Advanced Viewpoint, Cosimo Classics, 2007.

[266] P. A. Kidwell, A. Ackerberg-Hastings, and D. L. Roberts, Tools of American Mathematics Teaching, 1800–2000, Smithsonian Institution, 2008.

[267] R. Doerfler, The Lost Art of Nomography, The UMAP Journal **30** (2009), no. 4, 457–493.

[268] H. A. Evesham, The History and Development of Nomography, Docent Press, Boston, 2010.

[269] _____, The History and Development of Nomography, Docent Press, New York, 2010.

[270] A. Gluchoff, Artillerymen and mathematicians: Forest Ray Moulton and changes in American exterior ballistics, 1885–1934, Historia Mathematica **38** (2011), 506–547.

[271] J. Marasco, R. Doerfler, and L. Roschier, Doc, What Are My Chances?, The UMAP Journal **32** (2011), no. 4, 279–298.

[272] D. Zitarelli, The 1904 St. Louis Congress and Westward Expansion of American Mathematics, Notices of the American Mathematical Society **58** (2011), no. 8, 1100–1111.

[273] T. Bartlow, The Mathematical Life of Edward V. Huntington, Manuscript.

[274] R. Doerfler, Dead Reckonings; Lost Art in the Mathematical Sciences, Internet site.

[275] _____, A Zoomorphic Nomogram, Internet site.

[276] L. Roschier, PyNomo, Internet site.

[277] Letters exchanged by T. H. Gronwall and E. H. Moore, Eliakim Hastings Moore Papers, Box 1, Folder 25, The Joseph Regenstein Library, Department of Special Collections, The University of Chicago.

Index

Aberdeen Proving Grounds, 22, 25, 28, 33

Ballistics Technical Staff, 22, 30
Bennett, Albert A., 26
Bliss, Gilbert Ames, 15, 30
Bocher, Maxime, 25

Charbonnier, Paul, 33
Clark, J., 15, 36, 87, 104, 109, 122, 135, 139, 145

determinant equation, 77
Duporcq, Ernest, 87
Džems-Levy, G. E., 16

Eddy, Henry Turner, 9
Eichhorn, Melker Johann, 6, 13, 21
Epstein, L. Ivan, 114
equations
  class, 85, 106

Féjèr, Leopold, 13
Féjèr-Jackson-Gronwall Inequality, 14
Ford, Lester J., 25, 32

Gronwall Area Theorem, 14

Henderson, Lawrence Joseph, 20
Hewes, Lawrence I., 8, 17, 19, 32
Hezlet, Captain R. K., 2, 17
Hille, Einar, 12
homographic transformation, 37, 45, 82, 86
Huntington, Edward V., 9, 17, 32

index line, 76
isopleth, 76, 83, 114, 116, 132, 133

Jensen, J. L. W. V., 22

Kasner, Edward, 7, 20, 27, 32
Kellogg, Oliver D., 16, 19, 32
Klein, Felix, 8
Koebe, Paul, 12

Lalanne, Léon-Louis, 9
Lawrence Scientific School, 17
Lecornu, Joseph, 67
Lefschetz, Solomon, 13
Lipka, Joseph, 20, 23, 32

MacNeish, Harris, 32
Maschke, Heinrich, 13
Massau, Junius, 15, 67
Mittag-Leffler, Gösta, 12
Moore, C. L. E., 21
Moore, Eliakim Hastings, 5, 6, 8, 17, 22, 32
Morley, Frank, 6, 32
Moulton, Forest Ray, 24, 30, 31

nomogram
  acnodal, 142
  alignment, 1, 7, 19, 75
  circular, 83, 133
  collinear points, 35
  compound, 76
  conical, 36, 60, 88, 104, 107, 109, 135, 145
  crunodal, 144
  cubic, 145
  curved scales, 87
  cuspidal, 141
  family, 82, 106
  intersection, 1
  parallel scale, 79, 80, 87, 90
  parameters, 83, 106
  projective transformation, 82, 120, 133
  rectilinear, 45, 86, 88, 89

shared scale, 108, 145
single parameter, 109
single-curve, 88, 122
single-scale, 144
three rectilinear scales, 59, 88
three-scale, 125
three-variable, 75, 88
nomons, 20

Ocagne, Maurice d', 1, 14, 15, 18, 33, 35, 57, 109, 146
Office of Ordnance, 30

Peddle, John B., 9, 17, 18

Runge, Carl, 11, 18, 23

Saint-Robert, Paul de, 92
Sheffield Scientific School, 8, 17, 32
slide rule, 7, 76
Soreau, Rodolphe, 4, 15, 33, 146
Steinmetz, Charles P., 9
Strachan, R. C., 19, 22
summation property, 132

U.S. Naval Institute, 10
University of Chicago, 13, 24

Veblen, Oswald, 15, 23

Warmus, Mieczyså, 16
Weierstrass's elliptic curve, 113, 141
Weierstrass's elliptic function, 62, 113
Whittaker, Edmund, 20
Wiener, Norbert, 21

Printed in Germany
by Amazon Distribution
GmbH, Leipzig